Experiments in Reduced Gravity Sediment Settling on Mars

Experiments in Reduced Gravity Sediment Settling on Mars

Nikolaus Kuhn

ELSEVIER

AMSTERDAM • BOSTON • HEIDELBERG • LONDON
NEW YORK • OXFORD • PARIS • SAN DIEGO
SAN FRANCISCO • SINGAPORE • SYDNEY • TOKYO

Elsevier
Radarweg 29, PO Box 211, 1000 AE Amsterdam, Netherlands
The Boulevard, Langford Lane, Kidlington, Oxford OX5 1GB, UK
225 Wyman Street, Waltham, MA 02451, USA

Notices
Knowledge and best practice in this field are constantly changing. As new research and experience broaden our understanding, changes in research methods, professional practices, or medical treatment may become necessary.

Practitioners and researchers must always rely on their own experience and knowledge in evaluating and using any information, methods, compounds, or experiments described herein. In using such information or methods they should be mindful of their own safety and the safety of others, including parties for whom they have a professional responsibility.

To the fullest extent of the law, neither the Publisher nor the authors, contributors, or editors, assume any liability for any injury and/or damage to persons or property as a matter of products liability, negligence or otherwise, or from any use or operation of any methods, products, instructions, or ideas contained in the material herein.

British Library Cataloguing-in-Publication Data
A catalogue record for this book is available from the British Library.

Library of Congress Cataloging-in-Publication Data
A catalog record for this book is available from the Library of Congress.

ISBN: 978-0-12-799965-4

For information on all Elsevier publications
visit our website at http://store.elsevier.com/

This book has been manufactured using Print On Demand technology. Each copy is produced to order and is limited to black ink. The online version of this book will show color figures where appropriate.

TABLE OF CONTENTS

The final lines of this book were written on the first anniversary of the Mars Science Lab *Curiosity's* landing, 686 days on Earth or 666 sols on Mars. This year on Mars coincided with so far the most active part of our experiments on Martian surface processes. On August 6, 2012, our first reduced gravity flight for the *Mars Sedimentation Experiments* (**MarsSedEx**) was still about 3 months away and we had a very limited idea what to expect. At the day of the landing, the work I was doing was also very much down to Earth. After breakfast, I was anxiously watching CNN in a hotel in Ongwediva in northern Namibia where I was part of a team of researchers studying the livelihood of small-scale farmers. There we hoped to make a little contribution to both improving their crop yields as well as finding ways to use their land sustainably. This unquestionably more typical work for a geographer raises the question of how I ended up doing work on Mars and maybe even more importantly, why?

The answer to this question is straightforward: geography is driven to a large extent by curiosity and exploration of spaces, and for a trained geomorphologist, the surface of Mars offers a new world to explore. Having used experiments in much of my work, including the study of soil hydrology in northern Namibia, provided a tool for researching the Martian environment. Finally, understanding and studying Mars helps, in my mind, to learn about Earth. Taking an Environmental Systems Science perspective, one must always wonder why Mars is cold and dry now, but was once more habitable, and most importantly: why did Earth stay warm and wet for over three billion years and how stable this habitability actually is? Furthermore, in the context of studying and teaching, Mars also provides an opportunity to move outside the realm of conventional data collection and analysis, and therefore would allow students to be creative and think critically. If these reasons are not sufficient, the exciting scientific findings of the first year of the *Curiosity* mission (Life would have been possible!) and certainly the fascinating imagery of landscapes formed by the erosion of layers of sedimentary rocks might be an explanation in itself (at least to true geographers).

More immediately, the research on sedimentation on Mars was also driven by the question of how well the models we use on Earth to describe the processes that form these landscapes capture the effect of the different environmental conditions on our neighbor planet. A key driver in most models describing erosion, transport, and deposition of the eroded material is gravity. On Mars, it is reduced to 38% of the gravity we experience on Earth. As a consequence, for example, water flows much more slowly, achieving, in theory, a flow velocity of only 60% of the one it would achieve on Earth. The lower flow velocity reduces the kinetic energy of an identical mass of water moving downslope to, again in theory, approximately 33%. Apart from stretching the imagination, there are two drawbacks with these estimates: first, flowing water shapes the channel it is moving through, so calculations along the lines of "a given slope" or "a given stream channel" on Earth and Mars are not correct because there is a good chance that the relationship between form and process differs between the two planets. Second, most conventional models describing the relevant processes are highly empirical, i.e., based on observations made on Earth. The mathematical equations therefore often describe a process only within the boundary conditions of a terrestrial environment, raising questions about their applicability on Mars. From these considerations, which will be explained in more detail in Chapters 1 and 4 of this book, the question arose how sensitive the quality of the model output is to a change in gravity.

Discussions with colleagues on testing the quality of empirical models on flow hydraulics, erosion, and sediment movement quickly pointed toward an experimental approach onboard a reduced gravity plane. Experiments have a long tradition in geosciences and have also been used in planetary geomorphology. Some geomorphic research on mass movements had already been done on reduced gravity flights before, incidentally supporting our skepticism about semiempirical models. A cost analysis also showed that an experiment would be a more feasible way than numerical modeling based on first principles because the programming of a sophisticated computational fluid dynamics model takes a lot of time. Besides, wouldn't it be much better to actually see the sediment settling in reality than just on a computer screen?

The experimental approach chosen for this research reflects what experiments, and to a large extent, this book, are all about: measure something that cannot be calculated or monitored properly, or only with great effort.

Out of the range of processes that would be affected by gravity on Mars, settling velocity of sediment was selected because the results relate to many other processes and the way they are modeled. This is a further benefit of experiments: they can be designed to get a maximum number of answers, not hav"for".ing to submit to the constraints of the naturally occurring process domains. This book tries to illustrate the use and limits of experiments in geosciences. The reduced gravity conditions are both a stark contrast to Earth, enabling the identification of limits of existing models as well as the challenges of conducting a meaningful experiment on one of the recent and most exciting fields of geomorphic research: Mars.

Reporting ongoing research, some of the scientific results are preliminary and conclusions remain tentative. However, this serves the purpose of this book because the use of experiments is put into the larger context of Mars exploration. A further aim for setting a focus on experiments is to reach out to the wider geomorphic community, especially young researchers, and to share some conceptual thoughts on the purpose of experiments, their design, and the practical considerations that should be put into conducting them. This intention is reflected by the structure of this book. First, the need for experiments on Martian surface processes is explained, both from the perspective of current research questions as well as the quality of the models used to simulate these processes. A short overview of Mars follows to illustrate some of the major differences between Earth and Mars as well as our scientific interest. Moving toward experiments, the search for life on Mars is briefly presented, especially to document the cycle of scientific enquiry, swinging back and forth between observations, hypothesis, and the development and use of new research tools. Chapter 4 introduces sediment settling and the modeling of settling velocity as the main scientific theme of this book, followed by conceptual thoughts on experiments, the development of instruments for the measurement of settling velocity during reduced gravity flights of the MarsSedEx I and II missions. Chapters 7 and 8 give some practical advice on conducting the experiments themselves. Having young researchers in mind, the two chapters are also intended to introduce some critical thinking about preparing experiments in general. The following three chapters focus on the scientific outcome of the MarsSedEx I and II missions. The book concludes by putting the results of the missions in a preliminary perspective related to looking for traces of life on Mars and the further work that is required to improve our ability to model Martian surface processes properly.

Finally, my intention to conduct the work on Mars has a close relationship to other research and teaching I am involved with, such as the livelihood of Namibian farmers and the erosion of soil organic matter and its implications for climate change. Going to Mars tries to answer three questions: Where are we coming from? Where do we go to? And is or was there life on Mars? As a physical geographer, the first two questions fall into my area of professional interest, while the latter requires a good understanding of where to look for life or traces thereof (or spread life from Earth accidentally). Research on Mars, as this book, should have an impact on doing research and gaining understanding of Earth. By inviting to carry on, I would therefore ask the readers of this book to enjoy each critical review of models and their limitations and to be enthusiastic about making new discoveries, on Mars or in a farmhouse.

Movelier, June 2014

ACKNOWLEDGMENTS

I first and foremost would like to thank my wife Brigitte. She has been the strongest pillar of support throughout my career one can imagine. She also kept this project alive by developing the imaging techniques used to measure settling velocities, support the experiments, especially the research flights during her spare time, and endured this hobby project, both in our basement as well as during many weekends and evenings. The entire project, as well as the book, would have also been impossible without our Physical Geography team at the University of Basel, most notably Hans-Rudolf Rüegg for his enthusiastic and creative design and construction efforts. Further, the patient support of Ruth Strunk and Rosmarie Gisin for their boss going off to Mars receives my greatest gratitude. A special mentioning has to go to Marianne Caroni who left us much too early and whose supportive attitude to this project and her happy smiles are missed dearly. At the University of Basel, I would also like to thank Leena Baumann, Florence Bottin, Alexandra Diesslin, Lisette Kaufmann, and Dora Schweighoffer, for supporting the research, writing and illustrating effort for this book. A special mentioning also has to go to the Zero-G Corporation, especially Michelle Peters, for accommodating us and our weird apparatus onboard of their research flights. Among all the people directly or indirectly involved in the MarsSedEx missions, several need special mentioning: for the construction of the basic instrument carrying frame, I would like to thank Kai Wiedenhöft and Ernst Kuhn of the Mageba Textilmaschinen GmbH in Bernkastel-Kues, Germany; Dr. Martin Koziol of the Maarmuseum in Manderscheid, Germany, for providing the basalt sand samples; Dr. Jonathan Merrisson of the Department of Physics and Astronomy at the University of Aarhus, Denmark for enthusiastically supporting the idea of experiments in reduced gravity; and Dr. Andres Gartmann of the MCR Lab of the University of Basel for explaining CFD modeling to me. The funding for this project was provided by the University of Basel and Space Florida, which is gratefully acknowledged. The unwavering support of S., W., U., H., M., and L., as well as the understanding of our families is gratefully acknowledged. Finally, I would like to thank former Space Shuttle astronaut Bob Springer for being honest about taking motion sickness pills.

CHAPTER *1*

Sediment, Life, and Models on Mars

ABSTRACT

The search for life, or traces thereof, on Mars is one of the main drivers of current space exploration. Linked to this aim is the need for understanding the current and past Martian environment, both to reveal its past and present habitability in general, as well as to identify the sites where the preservation of traces of life is most likely. Sedimentary rocks are the prime targets for the ground exploration because they are proxies for the environmental history, i.e., habitability, of Mars and the most likely strata to preserve traces of life. The exploration of the sedimentary record relies on the ability to model surface processes, such as runoff, erosion, and deposition, on the Martian surface. These models support the selection of landing sites, ground targets, and the interpretation of the analysis conducted by rovers and landers. In this chapter, some initial results of the Mars Science Lab Curiosity's first year on Mars are summarized and put into the context of the search for life. This is followed by a critical review of models used to describe the relevant surface processes on Earth and their applicability to Mars.

1.1 SEDIMENTS AND LIFE ON MARS AT GALE CRATER

The Mars Science Lab Curiosity landed at Gale crater on August 6, 2012. The Gale crater landing site (Figure 1.1) was chosen because it appeared to offer a wide range of past aqueous and thus, potentially habitable environments, indicated by features such as outflow channels, an alluvial fan, as well as sequences of finely bedded deposits exposed in its center containing strata with phyllosilicates and sulfates (Figure 1.2).

Experiments in Reduced Gravity: Sediment Settling on Mars. DOI: 10.1016/B978-0-12-799965-4.00001-7

Fig. 1.1. Gale crater from orbit with landing ellipse for the Mars Science Lab Curiosity. Credit: NASA/JPL-Caltech/MSSS PIA 14290.

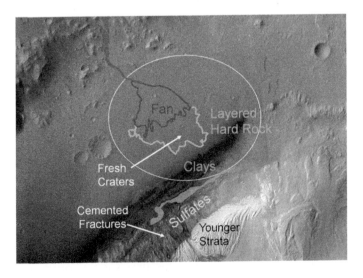

Fig. 1.2. Map of the relevant features for studying habitability in Gale crater. Credit: NASA/JPL-Caltech/MSSS PIA 14305. Sources: Reproduced with permission from Milliken et al. (2010) and Anderson and Bell (2010).

Fig. 1.3. *Curiosity on Mars, combined from images taken on sol 177 (February 3, 2013) and 270 of the mission.* Credit: NASA/JPL-Caltech/MSSS PIA 16937.

The landing site did not disappoint those who had selected it: more or less right from the point of touchdown, many features associated with deposition by water were encountered and sparked a flurry of data revealing the history and life-bearing potential of the analyzed sites (Figure 1.3). Right near the site where Curiosity touched down, called the *Bradbury Landing*, layers of sedimentary rocks were discovered on the first Martian day (sol) after landing. In the outcrop named *Goulburn*, a well-sorted gravel conglomerate was exposed by the blast

Fig. 1.4. Goulburn Scour, a conglomerate consisting of sand and pebbles blasted free by the engines during Curiosity's descent. The inset is magnified by a factor of two. The image was obtained by the Mast Camera (Mastcam) on August 19, 2012 (=sol 13, i.e., the 13th Martian day since Curiosity's landing). Credit: NASA/ JPL-Caltech/MSSS PIA 16187.

of the descent engine (Figure 1.4). Conglomerates are rocks consisting of a matrix of sand and embedded well-rounded pebbles. The pebbles at *Goulburn* were several centimeters in diameter and showed signs of orientation along the longitudinal axis. This type of deposition is indicative of sediment deposition from a flowing stream of water slowing down rapidly, possibly when leaving the confines of a channel. Further, similar outcrops were identified and analyzed by Curiosity during the first few weeks of the mission, which confirmed that the rover had landed on the remnants of an alluvial fan that had formed by erosion from the walls and deposition at their foot slopes in Gale crater. Based on the size of the pebbles embedded in the conglomerates and the slope of the alluvial fan, the flow velocity and depth of runoff from the crater wall could be estimated at 3–90 cm deep and flowing at a velocity of 2–75 cm/s.

The onward journey of Curiosity led to an area called *Yellowknife Bay* (Figure 1.5) with further sediments deposited by water (Figure 1.6). Here, three clearly different layers of sedimentary rock, called *Sheepbed, Gillespie Lake,* and *Glenelg,* were discovered. All three show clear evidence of being deposited by water, but differ in texture and structure. *Sheepbed* is a fine-grained mudstone consisting of particles smaller than 62.5 μm while *Gillespie Lake* and *Glenelg* are sandstones. *Glenelg* is also visibly

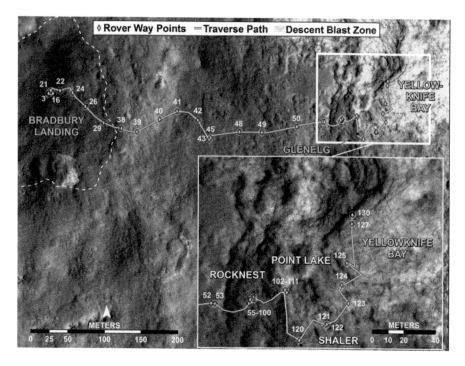

Fig. 1.5. Route of Curiosity during the first 130 days of surface operations. Credit: NASA/JPL-Caltech/MSSS PIA 16554b.

Fig. 1.6. Shaler outcrop in Yellowknife Bay recorded by the Mastcam on the 120th sol, December 7, 2012 on Earth, after landing Curiosity on Mars. The outcrop's patterned layers, called crossbedding, illustrate deposition by water. The rocks are part of the Glenelg formation. Credit: NASA/JPL-Caltech/MSSS PIA 16550.

cross-stratified, which is indicative of deposition from shallow-flowing water. The chemical composition of *Sheepbed* and *Gillespie* is similar to common Martian upper crust basalts, while *Glenelg*, on the other hand, is more alkaline. The chemical composition of all sediments indicates limited chemical weathering and thus, only a short exposure to a wet environment.

With regards to Curiosity's mission objectives, the *Sheepbed* sediments were most exciting because there is a strong evidence that they formed in a lacustrine, i.e., shoreline-type of environment that would offer habitability for microorganisms. This conclusion is based on a number of properties that can be used as proxies for the environmental conditions at the time of their formation. Sediment of a grain size similar to *Sheepbed* is abundant on Mars and mostly moved by wind. However, some features point toward a wet deposition of *Sheepbed*. The chemically relatively uniform layer of mudstone has a thickness of approximately 2 m. While the deposition of such a thick layer of dust is theoretically possible, it is unlikely to occur in a landscape that is shaped by fluvial processes at the same time. Dust deposition rates are low on Mars and it would take 10–20 million years to form such a thick layer of dust. Accumulation of a fine lake sediment layer of similar thickness, on the other hand, requires between 100s and 1000s of years. It is also very unlikely that such a thick dust layer has a chemical and mineralogical composition as uniform as the one observed in *Sheepbed* because its origin would most likely vary with the source, i.e., volcanic eruptions, each with a distinct chemistry. Further analysis of the *Sheepbed* sediments revealed that the sedimentary environment would have been habitable for chemolithotrophic microorganisms. The most important indicators for such habitability are the presence of hydrogen sulfides, which serve as an energy source and a pH in a near-neutral range. Furthermore, essential elements for life such as carbon, nitrogen, and phosphorus were detected in a form that would enable their use by microorganisms. Overall, the lake in which the *Sheepbed* sediments formed can therefore be considered as habitable for the type of microorganisms one would expect to have evolved on Mars. However, organic substances, such as methane, which are produced by living organisms, have only been found in traces lower than expected if life had been present. This leaves the Curiosity mission somewhere between a grand success because the potential habitability of Mars has been demonstrated, but the implicit hope of finding traces of life (although this was clearly not the scientific aim of the mission), has been disappointed.

The major lesson learned so far is that Mars is clearly more complex than expected. John P. Grotzinger, the lead scientist for the Curiosity Mission, summarized the findings by writing that "In this manner, the MSL mission has evolved from initially seeking to understand the habitability of ancient Mars to developing predictive models for the taphonomy of Martian organic matter".[1] Learning more about how to find these fossils is one goal the experiments described in this book aims to contribute to.

1.2 SEDIMENTATION AND THE TRACES OF LIFE ON MARS

Finding traces of life on Mars requires a good understanding where to look for them. The search for life in Mars sketched out in Chapter 3 of this book serves to illustrate this insight of 50 years of exploring Mars with landers and from orbit. This raises the question how taphonomy on Mars may differ from Earth. From a geomorphic perspective, the origin, transport, and especially deposition of sediment determine the likelihood of containing and preserving a fossil record. Consequently, predictive models for the taphonomy of Mars have to take into account the origin and development of the sediment.

The research done on the Curiosity data already revealed much information on the past environment of Mars, most notably the habitability at *Yellowknife Bay*. However, it became also clear that many of the rock formations were chemically largely unaltered from their volcanic origin and had just experienced what would appear to be physical weathering, erosion, transport, and deposition. Therefore, much of the information on the water that was involved in forming the rocks has to be unveiled from their structure and texture rather than their chemistry. This emphasizes the need for quantitative research on erosion and sedimentation on Mars (Figure 1.7).

There are two types of sediments, which may have preserved traces of life: first, biogenic sediments or mixtures of mineral and organic matter which have been transformed into rock immediately after deposition, e.g., in a lake, and not moved since; or second, deposits of organic material after erosion and transport. A further question to consider when looking for traces of life in sediment on Mars is "how wet" the past of the planet has been. On a warm and wet Mars, the likelihood of extensive lake or

---

[1] In Science 343, (2014) p. 387.

Fig. 1.7. *Interbedded sandstones, conglomerates, and a thin layer of fines deposited during the Triassic lower Buntsandstein formation, outcrop Fachhochschule near Trier, Germany. At the time, some 250 million years ago, the climate of the region was dry with irregular runoff and flooding. The sediments were deposited during such floods. The layer of fines may represent a pond or stillwater area where organic matter, either formed in situ in the water body or washed to this deposition point downslope. Would such a pattern of deposition be the needle to look for in the haystack of sediments on Mars?* Credit: N.J. Kuhn.

floodplain deposits rich in organic matter is far greater than on a desert planet with periodic rainfall and runoff forming ephemeral channels and short-lived playas. Finding traces of life on Mars, therefore, has to be based on a solid quantitative understanding of the movement and deposition of sediment in a range of environmental conditions. Using sediment as an environmental archive for the reconstruction of the presence and behavior of surface water is, consequently, as important as the search for traces of life itself for two reasons: first, the environmental history of Mars can be explored and second, looking for sediment layers potentially carrying traces of life is much more informed, especially when deciding on drilling into the ground that is not visible to cameras for a preliminary inspection. The interpretation of features in sedimentary rocks with regards to the environmental conditions at the time of their origin is an essential tool in geology. Outside Earth, principles of sedimentology were used in planetary research, most notably by the Apollo astronauts on the Moon and to interpret the imagery delivered by Mars probes, especially landers and rovers. However, key differences between the different planetary bodies exist. For Mars, with the strongest similarity in terms of landforms to Earth, the lower gravity strongly affects erosion, transport,

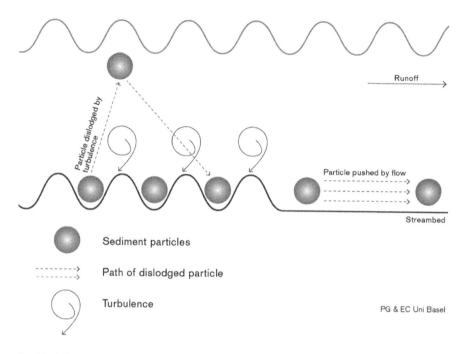

Fig. 1.8. Sediment movement in a river. The amount of sediment and the distance the particles move are determined by a range of factors: sediment has to be small and loose enough to be relocated by the water moving over the bed. At the same time, water has to be capable of moving the sediment. The transport distance depends on the height the particle is lifted to after dislodging and the flow velocity of the water. The different gravities on Mars and Earth affect almost all these factors for a given sediment and slope. On Mars, the flow velocity would be lower because gravity pulls the water with less force downslope. For a given stream channel this leads to greater runoff depth. Slower moving water has less kinetic energy to entrain a particle and lift it into the moving layer of water. However, the particle would also settle more slowly because gravity on Mars generates a smaller downward force. The settling velocity would still be proportionally greater than on Earth because the friction between particle and water depends on the settling velocity.

and deposition of sediment (Figure 1.8 and Figure 12.4). The effects are not straightforward, but act in a variety of ways on sediment.

1.3 PROCESS DESCRIPTION IN GEOMORPHIC MODELS

In the Earth and Planetary Sciences, geomorphologists study the processes shaping a landscape. Apart from analyzing the morphometry and composition of the surface and the measurement of the processes that shape it, numerical models are used to describe landform evolution or to predict future developments, such as the impact of changing climate. Models describing runoff, erosion, and transport on the surface of planetary bodies include those illustrated in Figures 1.8 and 12.4. Many of the processes are quantified on Earth using highly empirical and calibrated models. For the main theme of the book, particle settling, common modeling approaches are discussed in Chapter 4. Here and later on in Chapter 12, the modeling of flow velocity and critical shear stress for particle movement by running water, both processes closely related to sedimentation and habitability, are used to illustrate the potential shortcomings of applying semiempirical model working on Earth to Mars without testing their applicability. Using examples other than sediment settling also serves to show the wide range of processes where this critical review may be required. Runoff can be calculated by

$$q = w \times d \times v \qquad \text{Equation 1.1}$$

where

q = runoff rate
w = width of the channel
d = depth of the flow
v = flow velocity

For a given section of a channel, runoff varies little, but channel form can change and thus, affect flow velocity, width, and depth. Apart from width and depth, the roughness of the channel bed affects flow velocity and the kinetic energy of the flow entraining and transporting sediment. The effect of roughness on flow velocity can be estimated by empirically derived approaches such as the so-called Manning coefficient, named after the Irish engineer Robert Manning (1816–1897) who first developed numerical equations to assess the effect of roughness on flow velocity. Flow velocity is estimated by

$$v = \frac{k_n}{nR^{2/3}S^{1/2}} \qquad \text{Equation 1.2}$$

where

v = cross-sectional average velocity (m/s)
k_n = 1.0 for SI units
n = Manning coefficient of roughness
R = hydraulic radius (m) (=cross sectional area of flow (m²) divided by the wetted perimeter (m))
S = slope of channel (m/m)

For water flowing over natural surfaces, Manning coefficients range from 0.018 for smooth, fine-grained, and cohesive material to 0.023 for gravel and 0.035 for beds covered with large stones and cobbles. It is noteworthy that the effect of roughness on flow velocity varies with the hydraulic radius. Due to lower gravity on Mars, the same bed would therefore have a different influence on average flow velocity than on Earth. As a consequence, the interaction between water and bed is different.

Erosion, transport, and sedimentation are affected by these differences in flow hydraulics. Running water exerts a shear stress onto the bed (marked by the arrows in Figure 1.8). Sediments start moving when this stress is greater than the resistance of the sediment particles forming the bed (illustrated in Figure 12.4). For different types of bed material a so-called critical shear stress for particle motion, discussed in more detail in Chapter 12, can be observed. As for the Manning equation, the shear stress inducing motion is derived from observations with controlled runoff and is known as shear stresses. Figure 1.9 shows such an empirically derived relationship for shallow runoff. While the critical point for

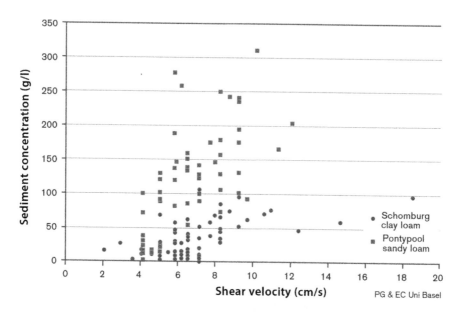

Fig. 1.9. Relationship between shear stress and sediment concentration observed during laboratory experiments on a 15-m flume in the Soil Erosion Laboratory, University of Toronto. The scatter highlights the variability of flow conditions, even under ideal lab conditions, and the limitations of empirically derived relationships on Earth.

so-called nonselective flow, i.e., with sediment particles of all sizes moving, is easy to identify, any further relationship for the differences between different size particles cannot be seen. The cloudy nature of the relationship between shear stress and sediment concentration also highlights the uncertainty of the output generated by the empirical models, even on Earth.

These common ways of modeling flow velocity and sediment movement, as well as their uncertainty illustrated in Figure 1.9, serve to illustrate the highly empirical nature of the current state of knowledge and practice with regards to sediment movement and sedimentation. A more detailed discussion of the specific limitations for the use of sediment settling velocity models on Mars is presented in Chapter 4. Most of these critical issues apply to all models used for erosion and sedimentation. So far, the relevance of gravity for these fluvial processes has largely been ignored, or assumed not to differ, in planetary research. However, experiments on the angle of repose of loose material during reduced gravity flights conducted by the group of Maarten Kleinhans from the University of Utrecht in The Netherlands raises some question. Their research has shown that the friction between still and moving particles differs significantly between Mars and Earth, with implications for the processes associated with the formation of particular landforms, for example, deltas and debris flow deposits. As a consequence of the critical differences identified by Maarten Kleinhans and his coauthors, one must question the use of semiempirical models to estimate runoff, erosion, and sedimentation on Mars. Consequently, the assessment of runoff on Mars and thus, implicitly the habitability of the Martian surface based on particle size measured from sedimentary strata should be done with utmost caution.

1.4 SEDIMENT TRANSPORT MODELS ON MARS

Achieving a good understanding of the environmental history of landing sites on Mars through the analysis of sedimentary rocks is required to learn about their habitability and the taphonomy of the potentially associated Martian organic matter. The Martian environmental history, including the current processes shaping the surface, is also of a more general interest than just identifying the likelihood of a sediment layer to contain and preserve traces of life for billions of years because by understanding the history of Mars, we learn about Earth. For both aims, models describing sediment movement on Mars are essential.

The strong empirical element in the modeling of flow hydraulics and sediment transport and the limitations of empirical models for reduced gravity identified by Maarten Kleinhans and coworkers raises the question whether the models developed for Earth work on Mars. John Grotzinger and his coauthors assume in their paper on *Sedimentary processes on Earth, Mars, Titan, and Venus* that "Empirical settling-velocity equations explicitly account for particle and fluid densities, fluid viscosity, and gravity, making them straightforward to apply to different fluids and planetary bodies".[2] The review above raises the question whether this assumption is actually true. While all relevant factors are integrated in common models used to describe runoff and sediment movement, they are often calibrated based on observations and associated with a coefficient, such as Manning's n, that fits them to the observed data. Often driving or controlling factors, such as slope and the hydraulic radius in the Manning equation, are modified by exponents without a physical explanation. The quality of such empirical relationships on Mars must be questioned because a factor such as gravity with a direct influence on depth and velocity of runoff, changes. Experiments testing and developing existing models relevant for flow hydraulics and sedimentation can therefore make a significant contribution to the ongoing research efforts on Mars and, thereby expand the limits of the models currently used. The major aim of this book is to illustrate how such testing was done during the Mars Sedimentation Experiments (MarsSedEx) project, with a focus on the reduced gravity flights of the MarsSedEx I and II missions. The settling of sediment particles was chosen because both, the way it is modeled and the process itself, relate well to other surface processes. Differences between Earth and Mars identified for settling velocity can therefore be used to review the quality of a wide

---

[2] Grotzinger, J.P., Hayes, A.G., Lamb, M.P., McLennan, S.M., 2013. Sedimentary processes on Earth, Mars, Titan, and Venus. In: Mackwell, S.J. et al. (Ed.), Comparative Climatology of Terrestrial Planets, University of Arizona, Tucson, pp. 439–472.

range of other processes models when applied on Mars. However, first the "boundary conditions" for this endeavor have to be set by giving a short introduction to the planet Mars and the history of the search for life on Mars.

BIBLIOGRAPHY

Anderson, R., 2010. Geologic mapping and characterization of Gale Crater and implications for its potential as a Mars Science Laboratory landing site. The Mars Journal 5, 76–128.

Golombek, M., et al. 2012. Selection of the Mars Science Laboratory Landing Site. Space Sci. Rev. 170, 97.

Grotzinger, J.P., 2014. Habitability, taphonomy, and the search for organic carbon on Mars. Science 343, 4, 387.

Grotzinger, J.P., Hayes, A.G., Lamb, M.P., McLennan, S.M., 2013. Sedimentary processes on Earth, Mars, Titan, and Venus. In: Mackwell, S.J. et al. (Ed.), Comparative Climatology of Terrestrial Planets. University of Arizona, Tucson, pp. 439–472.

Grotzinger, J.P., Sumner, D.Y., Kah, L.C., Stack, K., Gupta, S., Edgar, L., Rubin, D., Lewis, K., Schieber, J., Mangold, N., Milliken, R., Conrad, P.G., DesMarais, D., Farmer, J., Siebach, K., Calef, F., Hurowitz, J., McLennan, S.M., Ming, D., Vaniman, D., Crisp, J., Vasavada, A., Edgett, K.S., Malin, M., Blake, D., Gellert, R., Mahaffy, P., Wiens, R.C., Maurice, S., Grant, J.A., Wilson, S., Anderson, R.C., Beegle, L., Arvidson, R., Hallet, B., Sletten, R.S., Rice, M., Bell, J., Griffes, J., Ehlmann, B., Anderson, R.B., Bristow, T.F., Dietrich, W.E., Dromart, G., Eigenbrode, J., Fraeman, A., Hardgrove, C., Herkenhoff, K., Jandura, L., Kocurek, G., Lee, S., Leshin, L.A., Leveille, R., Limonadi, D., Maki, J., McCloskey, S., Meyer, M., Minitti, M., Newsom, H., Oehler, D., Okon, A., Palucis, M., Parker, T., Rowland, S., Schmidt, M., Squyres, S., Steele, A., Stolper, E., Summons, R., Treiman, A., Williams, R., Yingst, A., Team, M.S., 2014. A habitable fluvio-lacustrine environment at Yellowknife Bay, Gale Crater, Mars. Science 343, 14.

Hand, E., 2011. NASA picks Mars landing site. Curiosity rover will explore Gale Crater, which may hold clues to past habitability. Nature 475, 1.

Julien, P.Y., 2010. Erosion and Sedimentation, second ed. Cambridge University Press, Cambridge, p. 371.

Kleinhans, M.G., Markies, H., de Vet, S.J., in 't Veld, A.C., Postema, F.N., 2011. Static and dynamic angles of repose in loose granular materials under reduced gravity. J. Geophys. Res. 116, E11004.

Knighton, D., 2014. Fluvial Forms and Processes: A New Perspective, second ed. Routledge, New York, p. 383.

Komar, P.D., 1979. Modes of sediment transport in channelized water flows with ramifications to the erosion of the Martian outflow channels. Icarus 42, 13.

Malin, M.C., Edgett, K.S., 2003. Evidence for persistent flow and aqueous sedimentation on early Mars. Science 302, 4.

McLennan, S.M., Anderson, R.B., Bell, J.F., Bridges, J.C., Calef, F., Campbell, J.L., Clark, B.C., Clegg, S., Conrad, P., Cousin, A., Des Marais, D.J., Dromart, G., Dyar, M.D., Edgar, L.A., Ehlmann, B.L., Fabre, C., Forni, O., Gasnault, O., Gellert, R., Gordon, S., Grant, J.A., Grotzinger, J.P., Gupta, S., Herkenhoff, K.E., Hurowitz, J.A., King, P.L., Le Mouélic, S., Leshin, L.A., Léveillé, R., Lewis, K.W., Mangold, N., Maurice, S., Ming, D.W., Morris, R.V., Nachon, M., Newsom, H.E., Ollila, A.M., Perrett, G.M., Rice, M.S., Schmidt, M.E., Schwenzer, S.P., Stack, K., Stolper, E.M., Sumner, D.Y., Treiman, A.H., VanBommel, S., Vaniman, D.T., Vasavada, A., Wiens, R.C., Yingst, R.A., Team, M.S., 2014. Elemental geochemistry of sedimentary rocks at Yellowknife Bay, Gale Crater. Mars. Sci. 343, 10.

Milliken, R.E., Grotzinger, J.P., Thomson, B.J., 2010. Paleoclimate of Mars as captured by the stratigraphic record in Gale Crater. Geophysical Research Letters 37.

Wagner, H.W., Kremb-Wagner, F., Koziol, M., Negendank, J.F.W., 2012. Trier und Umgebung. 3. völlig neu bearbeitete Auflage. Gebr. Borntraeger, Stuttgart, p. 396.

Williams, R.M.E., Grotzinger, J.P., Dietrich, W.E., Gupta, S., Sumner, D.Y., Wiens, R.C., Mangold, N., Malin, M.C., Edgett, K.S., Maurice, S., Forni, O., Gasnault, O., Ollila, A., Newsom, H.E., Dromart, G., Palucis, M.C., Yingst, R.A., Anderson, R.B., Herkenhoff, K.E., Le Mouélic, S., Goetz, W., Madsen, M.B., Koefoed, A., Jensen, J.K., Bridges, J.C., Schwenzer, S.P., Lewis, K.W., Stack, K.M., Rubin, D., Kah, L.C., Bell, J.F., Farmer, J.D., Sullivan, R., Van Beek, T., Blaney, D.L., Pariser, O., Deen, R.G., Team, M.S., 2013. Martian Fluvial Conglomerates at Gale Crater. Science 340, 1068–1072.

INTERNET RESOURCES

http://exploration.esa.int/mars/

http://insight.jpl.nasa.gov/home.cfm

http://photojournal.jpl.nasa.gov

CHAPTER 2

Overview of Mars

ABSTRACT

Mars and Earth have many similarities, but are also marked by the stark difference in their current environmental conditions. The shared "wet and warm" early history raises the question why both planets took a different development. This in itself is a worthwhile research topic, helping to understand Earth, but it is also essential to guide the search for traces of life on Mars. In this chapter, a short overview of Mars and its past and current environment is given. This overview sets the stage for the following chapters because this brief look at Mars as a planet is a prerequisite for understanding the current strategies of exploration and the challenges faced when selecting landing sites and analyzing sedimentary rocks and landforms. Based on a comparison of Mars and Earth and the geologic history of Mars, the relevance of researching Mars is also discussed.

2.1 MARS AND EARTH

Mars is the fourth and outermost terrestrial planet of our solar system, orbiting the Sun at an average distance of 228 million kilometers (Figure 2.1). Craters, volcanoes, and fluvial channels generate a landscape that resembles a mixture between the Moon and deserts on Earth.

Further similarities include the seasonality of the climate. Despite the greater distance from the Sun than Earth, Mars can still be considered to be in the habitable zone, defined as warm enough for water to occur as a liquid, of the solar system. The solar energy received by Mars at the top of its atmosphere at the equator corresponds roughly

Experiments in Reduced Gravity: Sediment Settling on Mars. DOI: 10.1016/B978-0-12-799965-4.00002-9

Fig. 2.1. Mars seen by the Hubble Space Telescope. A view similar to this one offered the most information on Mars that could be achieved before space probes reached the planet. hs-2005-34-i-full_tif.tif. Credit: NASA, ESA, The Hubble Heritage Team (STScI/AURA), J. Bell (Cornell University), and M. Wolff (Space Science Institute).

to the amount received on Earth at 64.5° latitude (Table 2.1). Still, today, Mars is a cold planet with a thin atmosphere, which may permit liquid water at best only for few hours and few days per year at sites with ideal temperature conditions and a supply of ground ice or snow (Figures 2.2 and 2.3). Furthermore, while the solar energy received by Mars is lower than at Earth's distance from the Sun, the protection of the surface from ultraviolet and other radiations is also much reduced because of the thin atmosphere and largely missing magnetic field.

Property	Mars	Mars/Earth
Table 2.1. Key Properties of Earth and Mars Compared		
Mass (10^{23} kg)	6.42	0.107
Equatorial radius (km)	3397	0.533
Mean density (g/cm³)	3.93	0.713
Equatorial gravity (m/s²)	3.706	0.377
Solar irradiance (W/m²)	589.2	0.409
Length of day (1 sol)	24 h, 37 min, 22 s	1.029
Axial tilt	23°59'	1.074
Orbital eccentricity	0.093	5.593
Length of year days	686.98	1.881
Average distance from Sun (10^6 km)	227.92	1.524
Average surface temperature	214 k/ −59 °C	0.74
Atmospheric pressure (hPa)	6.1	0.006
Carbon dioxide in the atmosphere (%)	95.32	0.0004
Oxygen in the atmosphere (%)	0.13	0.006
Nitrogen (%)	2.7	0.035
Argon (%)	1.6	1.71

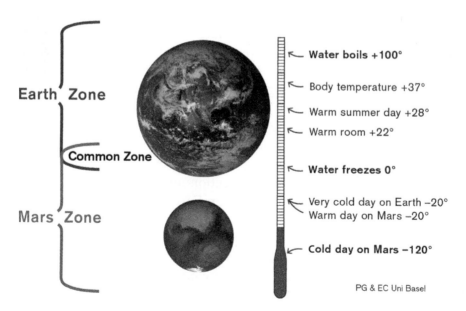

Fig. 2.2. Habitable temperature zones on Mars and Earth defined by the presence of liquid water. Currently, Mars achieves those temperatures only rarely.

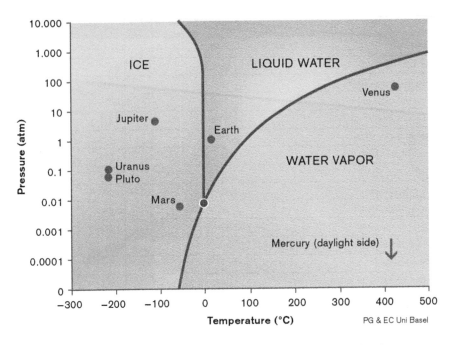

Fig. 2.3. Water–temperature and pressure triangle, indicating that even if temperatures are above freezing point, low pressures may prevent liquid water to be present, by causing an immediate evaporation into the thin atmosphere.

Despite these differences, astronomers, and also geoscientists and biologists have a strong interest in Mars. The reason for this interest lies both in the similarities between Earth and Mars as well as their differences. While landforms on Mars appear familiar, they are much older than on Earth and originate from a different set of interaction between controlling factors and processes. The two key differences between Mars and Earth are the lack of plate tectonics (Figure 2.4) and the very limited set of surface processes, which also sculpt the Martian surface at much lower rates than the terrestrial counterparts. No plate tectonics avoids the destruction and structural changes of crusts, common at plate boundaries on Earth. As a consequence, the surface of Mars is on average much older than on Earth, generally on the order of billions of years rather than tens to hundreds of million years. The lack of a thick atmosphere with rainfall contributes to this preservation of surface features because both weathering and erosion have much lower rates than on Earth. Therefore, the Martian topography shows a mixture of features that documents more or less the entire morphogenetic history of Mars.

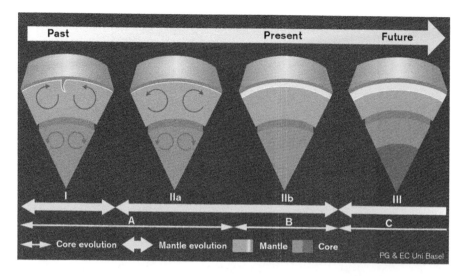

Fig. 2.4. The interior of Mars through time: on a young Mars, mantle convection, like still on Earth, may have been present, but rapid cooling and heat transfer caused the mantle movement to cease and thus stopped plate tectonics. Based on Forget et al. (2008).

On a global scale, Martian topography is characterized by the contrast between the highlands of the southern and the lower lying, smooth basins of the northern hemisphere (Figure 2.5). This difference, known as the Martian dichotomy, is assumed to have formed early during the history of Mars. A fully tested hypothesis explaining the dichotomy has not been presented yet. Most recent studies point toward the combined effect of large impacts and an uneven mantle convection on uplift and volcanism on the southern hemisphere. As a consequence, the crust of the northern hemisphere is thinner (30–35 km in the North compared to 45–58 in the South) and the elevation is approximately 3 km lower than in the South.

2.2 GEOLOGIC HISTORY OF MARS

Based on the density of impact craters, the geologic history of Mars can be divided into three periods, each named large-scale surface features on Mars formed during these periods (Figure 2.6). The oldest period is referred to as Pre-Noachian and covers the time from the accretion from the protoplanetary disk surrounding the Sun until the differentiation of the planet into crust, mantle, and core. This period began at the same time as the formation of Earth some 4.5 billion years ago and ended

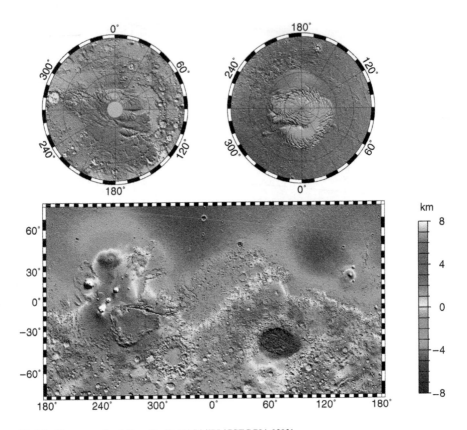

Fig. 2.5. The topography of Mars. Credit: NASA/JPL/GSFC PIA 02031.

approximately 4.1 billion years ago. On Mars, the end is marked by the formation of the Hellas impact basin. The next period is called the Noachian after Noachis Terra, west of the Hellas impact basin. During the Noachian, Mars experienced its most active formation of surface features through both impact cratering as well as the formation of river valleys. This period, which ended approximately 3.7 billion years ago, may have also experienced the presence of large lakes and a northern ocean on Mars. Between 3.7 and 3.1 billion years before present, widespread volcanic eruptions caused the formation of large lave plains, such as Hesperia Planum, after which the Hesperian period is named. Linked to the volcanism are catastrophic releases of water, causing so called megafloods, which left large outflow channels, especially at the border of the Martian dichotomy. During the Hesperian, ephemeral water bodies formed in the lower lying areas north of the equator. The final major

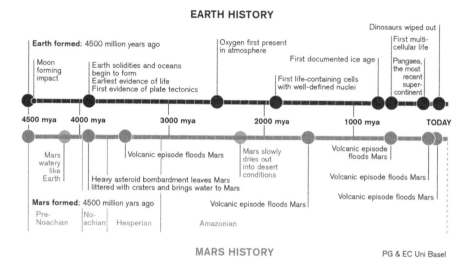

Fig. 2.6. Geologic history of Mars and Earth. Based on Forget et al. (2008).

period of Martian history is called the Amazonian after the Amazonis Planitia, west of Olympus Mons. During this period, which lasts to the present, landforms changed due to limited and small impact cratering, lava flows, water, ice- and frost-related surface processes as well as wind erosion and deposition. It is noteworthy that on Earth only a fraction of the terrestrial surface, the old shields, have an age of more than a billion years, while 80% is younger than 500 million years. This difference in age illustrates that the surface of Mars is old, by comparison, and thus preserved a young stage of planetary formation.

2.3 RESEARCHING MARS

The old surface of Mars offers a glimpse into the past of terrestrial planets and thereby enables basic research in the life and environmental sciences. The three questions driving current planetary science can be summarized as:

1. Where are we coming from?
2. Where are we going?
3. Is there life on other planets?

Each of these questions has major implications for our under-standing of life on Earth. Understanding the origin of Earth and in particular the interaction of factors that created and maintained a habitable planet is of great significance at what is considered the change from a quasi-naturally functioning planet to one that is managed by humans.

During the past 700 million years, the environment on Earth has experienced approximately five catastrophic changes that led to mass extinctions, but may have also caused a push for the evolution of new species. Most notably, the impact of a large meteorite at the end of the Cretaceous period some 60 million years ago is considered to be a major factor for the cooling of the climate which contributed to the extinction of the dominating dinosaurs, as well as the subsequent rise of the mammals. External factors, such as meteorites, are not required for catastrophic change. About 700 million years ago climate was much cooler and large parts of the planet covered with ice and snow. Some scientists refer to this period as "Snowball Earth" and geologic records suggest that much of the life disappeared due to the lack of habitat, now frozen. The cooling is attributed to tectonic activities, similar to another mass extinction about 200 million years ago. Once over, the Cambrian explosion of multicellular life began 541 million years ago. Even earlier, the evolution of life itself had almost led to its destruction: about 2.5 billion years ago, cyanobacteria released oxygen through photosynthesis. Free oxygen in the atmosphere, and the increasing concentrations in water, are toxic for anaerobic organisms and may have almost destroyed life on Earth.

The lack of oxygen in the Martian atmosphere was seen as a sign of limited or even no presence of life by James Lovelock and started his thoughts on Earth as a self-regulating planetary system, maintaining climate suitable for life. His thoughts evolved into the so-called Gaia theory and illustrate the need for comparative planetary studies. Comparing the planetary history of Mars and Earth sheds a light on the stability of the surface conditions of a terrestrial planet and the mechanisms that control them, basically, does a Gaia hypothesis-like self-regulating mechanism exist or were we just lucky to get thus far?

Understanding the interaction within the Mars and Earth planetary systems also helps to assess the potential risks of human-induced environmental change on the life-supporting conditions on Earth in the

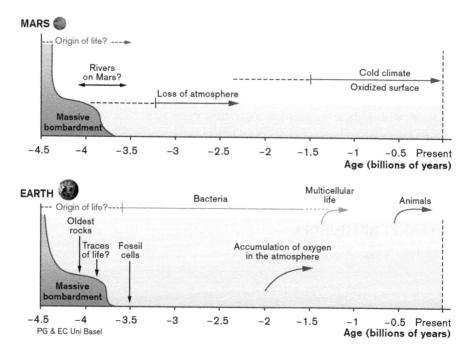

Fig. 2.7. Planetary history and life on Mars and Earth. Based on Forget et al. (2008).

future (Figure 2.7). Finally, Mars appears to have offered habitability, at least by the standards of early microorganism on Earth for a time period that was sufficient for life to evolve on Earth. Finding out whether a similar process occurred on Mars (or not) is therefore not only of scientific interest, but has even a philosophical importance with regards to the likelihood of being (more or less) alone in the universe. Answering each of those questions relies heavily on understanding the environmental history of Mars. Sediments as those examined by Curiosity and its predecessors are archives and understanding them is the key to unraveling the information they contain. The work presented in this book aims at making a small contribution to deciphering what those archives can tell us.

BIBLIOGRAPHY

Balme, M.R., Bargery, A.S., Gallagher, C.J., Gupta, S. (Eds.), 2011. Martian Geomorphology. Geological Society, Special Publications, London, p. 356.

Barlow, N.G. (Ed.), 2008. Mars. An Introduction to its Interior, Surface and Atmosphere. Cambridge University Press, New York, p. 264.

Bell, J. (Ed.), 2008. The Martian Surface. Composition, Mineralogy, and Physical Properties. Cambridge University Press, New York, p. 652.

Carr, M.H. (Ed.), 2006. The Surface of Mars. Cambridge University Press, New York, p. 322.

Golabek, G.J., Keller, T., Gerya, T.V., Zhu, G., Tackley, P.J., Connolly, J.A.D., 2011. Origin of the Martian dichotomy and Tharsis from a giant impact causing massive magmatism. Icarus 215, 346–357.

Forget, F., Costartarde, F., Lognonné, P. (Eds.), 2008. Planet Mars: Story of Another World, Springer-Praxis Books in Popular Astronomy. Springer in association with Praxis Pub., Berlin; New York; Chichester, UK.

Squyres, S., 2006. Roving Mars: Spirit, Opportunity, and the Exploration of the Red Planet. Hyperion, New York, p. 432.

INTERNET RESOURCES

http://photojournal.jpl.nasa.gov

Search for Life on Mars

ABSTRACT

The search for life on Mars is intrinsically linked to the state of knowledge about our neighboring planet. An interesting circle appears to be closed at the moment, from the steadfast assumption that there is naturally life on other planets, to the wishful thinking that there was, to the increasing disappointment when scientific evidence mounted against life on Mars, to finally showing that at least it would have been habitable for primitive forms of life known from Earth. The limbo between being potentially habitable and not having revealed traces of life (yet) challenges scientists to look for life in the right place. Finding this place requires a good understanding of the surface processes that shaped the Martian surface, both in the past and at present. This short introduction to the search for life on Mars prepares the main theme of the book by giving an insight in how scientific evidence, limited by distance and technology, has been used to search for life (or the lack thereof) on Mars. The chapter also illustrates the need for studying surface processes on Mars to make the most use of the available and planned space exploration hardware searching for life.

3.1 PRESPACE AGE RESEARCH

The scientific exploration of Mars can be closely linked to the development of technology that enabled the observation of features on the Martian surface. In the seventeenth century, astronomers such as Christian Huygens and Giovanni Cassini were able to measure Martian

Experiments in Reduced Gravity: Sediment Settling on Mars. DOI: 10.1016/B978-0-12-799965-4.00003-0

orbital parameters and noted changing markings on the surface of Mars. Some 100 years later, William Herschel identified polar ice caps and likened them to those on Earth. This led to the conclusion that Mars has an atmosphere with properties similar to those on Earth. While this conclusion appears very brave today, it was reasonable with the information available at the time. Interestingly, the early atmospheres of Earth, Mars, and Venus are actually assumed to be relatively similar, just taking different developments later on during planetary evolution. The notion of Mars teeming with life received more support during the nineteenth and early twentieth century. Dark markings moving between poles and equator in a seasonal pattern were interpreted as the spread of vegetation during spring and summer, followed by retreat in the winter. The French astronomer Georges Fournier wrote in 1909 that "As the spring advances, the dark shading progressively encroaches from the pole towards the equator across the seas".[1] Spectral analysis of the light received from Mars, however, revealed that the dark shading did not contain chlorophyll and was thus, not caused by the spread of a lush vegetation cover. While cacti or lichens were considered initially, nonbiological causes were proposed, such as moisture changes, ash, or dust (Figure 3.1).

The advancement of knowledge at the beginning of the twentieth century would have probably laid the earlier ideas of a living Mars to rest had not two astronomers, unintentionally, contributed and developed an intriguing hypothesis claiming highly developed life must exist on Mars. The two astronomers were Giovanni Schiaparelli and Percival Lowell. The former observed what he called "canali" on the surface of Mars in 1877. He used the term simply to describe a linear feature on the surface. Lowell, inspired by the work of Schiaparelli, devoted his time and wealth (including the construction of one of the finest observatories at the time in Flagstaff, Arizona) to the study of Mars and published two landmark books in 1906 and 1909 identifying the "canali" as canals constructed by a well-organized Martian society to redistribute the water on Mars. Lowell, by no means an amateur scientist, was also an accomplished writer and could convincingly, at least for most outsiders, publish his assessment of the observations he and the professional astronomers working with him made. He offered explanations to

---

[1] In Taylor, F.W. The Scientific Exploration of Mars, p. 8.

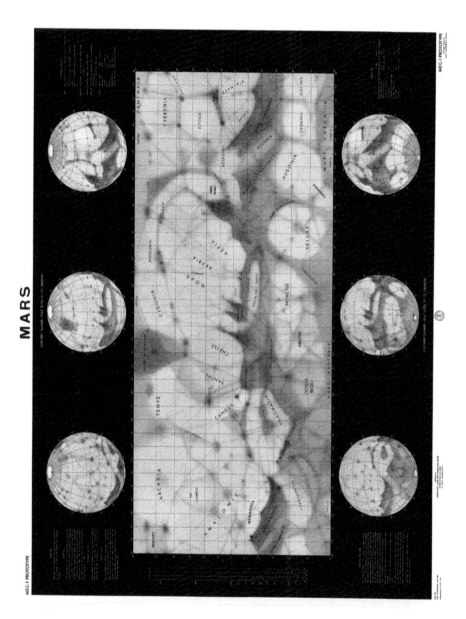

Fig. 3.1. Earl Slipher's 1960 Mars map: no more signs of life? Source: International Planetary Cartography Database, http://planetologia.elte.hu/ipcd/mec-1.jpg

questions raised about his claim. For example, a canal, even if several kilometers wide, could not be seen with the telescopes from Earth at the time. Lowell explained the apparently much wider structures on Mars by assuming that he saw not just the canal, but also a band of irrigated land surrounding it. Critics at the time, such as French astronomer Eugene Antoniadi struggled to see the features Lowell described. Further, the English naturalist Alfred Russel Wallace, summarizing observations by other astronomers, pointed out that due to the distance between Mars and the Sun the planet was likely much too cold and that the atmosphere was too thin to support vegetation, let alone humanoid beings. Over the following decades, an increasing and improving number of observations did not support Lowell's ideas. It turned out that the canals he observed were probably a combination of an optical and neurological illusion. The final judgment was made when the first images of Mars were transmitted to Earth by Mariner 4 in 1965, revealing a dry and moon-like surface of the southern highlands (Figure 3.2).

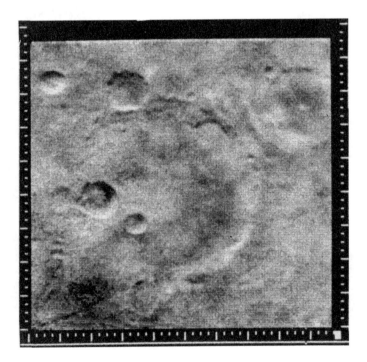

Fig. 3.2. Mariner 4 image of Mars showing the Mariner craters. The landscape seen on the first images transmitted by spacecraft flying by Mars showed a cratered landscape, similar to the Moon with few features resembling a terrestrial surface. Credit: NASA Image, ID number Mariner 4, frame 11E, taken on July 15, 1965.

3.2 LOOKING FOR LIFE ON MARS

The observations made by the Mariner probes were initially very disappointing for those who hoped to find evidence of habitability on Mars and may be, even wide-spread presence of microorganisms. The planet looked cold, dry, and much more like the Moon than Earth. To a certain extent, this first close-up impression of Mars was a coincidence because the probes flying by (Mariner 4, 6, and 7) took images of the southern hemisphere of the planet. Only with orbiting probes, covering most of the planet's surface, the dichotomy of the topography and the large-scale fluvial features at the border between high and lowlands were discovered. The apparent presence of water renewed the interest in Mars and contributed to the development of the Viking program in the 1970s. The Viking probes, reaching Mars in 1976, consisted of an orbiter and a lander (Figure 3.3). Their scientific aims were to map the Martian surface based on high-resolution images, to characterize the atmosphere and climate and to search for evidence of life. The search for life used four sets of instruments installed in the landers. Unlike Curiosity which "just" explores the habitability of Mars, planetary scientists were far

Fig. 3.3. Image taken by the Viking 2 Lander image taken on November 2, 1976, showing the spacecraft and part of Utopia Planitia, looking due south. The American flags, color grid, and bicentennial symbols on the spacecraft were used for color balance. Credit: NASA, Image ID number: 21C056.

more optimistic to actually find life in the 1970s. Consequently, the experiments aimed at looking for traces of current microbial metabolism rather than just its leftovers from billions of years ago.

The Pyrolitic Release (PR) experiment was designed to detect traces of substances generated by photosynthesis. Samples of Martian regolith were incubated with light, water, and carbon monoxide and carbon dioxide enriched in C^{14}. After several days of incubation, the radioactivity of the solid regolith sample was analyzed. An increased radioactivity would have been indicative of microbial activity, incorporating the carbon isotopes into biomass. However, the experiment returned no result. To identify existing organic substances, a combined gas chromatograph and mass spectrometer experiment was part of the instrument suite onboard the Viking landers. Surface samples are heated and the emitted gases can be analyzed for their chemical composition. As for the PR experiment, no positive results were returned. The gas exchange experiment combined incubation and gas chromatography. Samples of surface material were inoculated with organic and inorganic nutrients as well as water. Any microorganisms present in the sample were expected to convert some of the nutrients into gases that could be detected with the gas chromatograph. But again, no positive results were returned. The final and to date most controversially discussed experiment is the Labeled Release experiment. Turning the principle of the PR experiment around, the Labeled Release experiment aimed at identifying the release of carbon isotopes from regolith samples inoculated with a water and nutrient solution containing C^{14}. This experiment did generate a positive, but controversial, result. While C^{14} release was seen as evidence for microbial activity, an expected increase in release with ongoing addition of nutrients was not observed. Such a reaction would be expected if the increasing supply of nutrients activates a growing number of microorganisms. A declining release, on the other hand, is evidence for an inorganic chemical reaction that achieves a new equilibrium. The results of the Labeled Release experiment were considered to be inconclusive by principal investigator, Gilbert Levin, in the 1970s. Experiments on the behavior of organic substances under the extreme temperature and radiation conditions on Mars conducted on Earth showed that inorganic reaction may have also led to the release of labeled carbon. As a consequence, most scientists consider the Viking search for life as negative.

However, the detection of perchlorate by the Phoenix lander in 2008 offers the possibility that some gases were destroyed before being detected. Furthermore, the sensors onboard Viking may have not been able to detect low amounts of gases, thus missing part of the release. Gilbert Levin and coworkers still maintain that the results of the Labeled Release experiment may contain some evidence for microbial activity on Mars. Using advanced statistical cluster analysis, they argue as recently as 2012 that the temporal patterns of gas release from the Viking experiment differs from what would be expected by an inorganic reaction. Needless to say, this latest interpretation is also controversial.

3.3 CURRENT STRATEGIES FOR MARS EXPLORATION

Today, the Viking experiments may look naïve, explaining observations in a context that assumes that life is present, may be still under the influence by the enthusiasm of Percival Lowell. The negative or at best inconclusive results of the Viking missions left our understanding of life on Mars in a limbo, almost a déjà vu of the outcome of previous advances in knowledge. As a consequence, Mars exploration almost stopped for two decades (see a list of all missions to Mars, their objectives, and fate in Appendix I). Still, based on the results of Viking and its predecessors, our understanding of Mars grew. At the same time, the complexity of Earth as a system of interacting spheres evolved. The new Earth Systems Science contributed to the recognition that Mars was an equally complex system worth studying to learn both about the evolution of life as well as the fate of our own planet Earth.

The vast evidence for water on Mars and the similarity of young Earth and Mars provided therefore enough encouragement to renew the exploration of Mars. This research strategy, developed under the auspices of the National Aeronautics and Space Administration (NASA), was adopted by most space agencies. With regards to the search for life, the strategy involves three consecutive stages. The first stage, beginning with the Mars Global Surveyor and Mars Pathfinder, both launched by NASA in 1996, summed up under the heading "Follow the Water," aimed at systematically exploring the past and present presence of liquid water at or near the surface of Mars. Until now, all missions to Mars either focused or had an element of increasing our understanding of water on Mars. It became also clear that putting an immobile lander where technically possible, might not be the best way to hit a spot teeming with (traces of) life on a cold planet fried by shortwave radiation. Based on this insight, the recent era of Mars exploration began some 25 years ago, putting Mars Pathfinder and its rover Sojourner onto Mars in 1996. Furthermore, exploration moved to areas outside the safer landing zones of the planitias to the north (Phoenix) and closer to the border areas of the Martian dichotomy (Mars Exploration Rovers Spirit and Opportunity) and even into Gale crater, getting as close as possible to outcrops revealing Mars' past. With Phoenix, the suite of instruments carried by the landers and rovers was also expanded to follow the second and third parts of the exploration strategy. These involve the assessment of the habitability of Mars as well as the detection of traces of life. Unlike Viking, no microbial activity is induced and measured, but organic substances likely associated with past life are sought. At the same time, mapping of the planet and the study of the atmosphere of Mars are advanced. The next major step in the search for life on Mars is the sampling and analysis of material below the surface, where organic substances are protected from the destructive exposure to ultraviolet and other radiations from space. Both NASA's 2016 InSight Lander and ESA 2018 ExoMars rover carry drilling tools to achieve this objective (Figure 3.4). Interestingly, their search for life on Mars turns away from sites that might be considered habitable at present. The main reason is the risk of contaminating the sites, or even the planet, with terrestrial microorganisms. To prevent contamination the Viking landers and shells were placed in a pressurized chamber and sterilized by exposing them to a temperature of 111°C for 40 h. While such sterilization of landers is possible, it is not feasible. Furthermore, searching for traces of life offers a much wider study area, and thus, simpler landing and movement, than looking for life in currently potentially habitable sites.

3.4 LOOKING FOR SITES CONTAINING TRACES OF LIFE

The size of a landing site on Mars varies with the technology used by the lander, but generally varies between a 10 and 100 km long and 2 and 20 km wide ellipse. Further constraints include the latitude, elevation, and topography of the terrain. In addition, rovers are moving only slowly, the Mars Exploration Rover Opportunity managed 39 km

Fig. 3.4. The European Space Agency's ExoMars Rover with 2 m drill, exploring the near surface environment protected from radiation and thus, preserving traces of life. With a mission lasting only 7 months and few opportunities to drill, the selection of the right sediment is essential for success. Credit: ESA/AOS Medialab.

in 10 years. This distance may seem fair compared to the size of the landing ellipse, but it should be noted that Opportunity has exceeded its life expectancy by a factor of 5 to 10, and even 40 times considering the 90 days of the planned mission. The limited mobility of rovers and landers therefore requires a careful selection of landing sites based on their landforms, lithology, and history.

Landing sites have to combine a likely habitability in the past, chance of accumulation of traces of life, their preservation during diagenesis and afterwards, followed by excavation in recent past without destruction by radiation. The multitude of criteria raises several questions

Fig. 3.5. Hypanis Vallis and an eroding alluvial fan or delta. The next place to look for life on Mars? The images taken from orbit show a promising combination of features: at the bottom center, left, a channel is visible whose sediments were deposited in an alluvial fan or delta seen in the center top right of the image. Craters in these surfaces may offer outcrops, as well as scouring by wind. Source: http://themis.asu.edu/node/5387

related to landforms and their lithological properties. Candidate surfaces could have formed at sites with in situ life, such as soils, lakes, riverine wetlands, and wet near surface environments. Such sites can be identified from orbit based on their mineralogy and topography (Figure 3.5). Most promising are areas with high concentrations of clays, which may have formed under the influence of water. Clays are also good at preserving organic substances. However, aiming just for "clay" bears several risks. No Mars analog site on Earth is clay rich, so the association of clays and microorganisms is unclear. High concentrations of clay do not work well as a habitat on Earth because they have small and discontinuous pores, which limit the movement of water and air. Only bioturbation by roots, large animals, or tillage operations by humans, can create an environment suitable for microorganisms. Finally, large clay deposits may have formed on Mars from ash that was deposited during the eruption of a volcano. The chemistry and mineralogy of these deposits can be similar to those deposited in water. Looking for further evidence that water was present in the past, such as layered sediments or fluvial features, therefore reduces the risk of landing in a dead and dry place.

A second type of target site is sediments that contain traces of life, but may have been moved by subsequent erosion and deposition. Such deposition may generate a concentration of organics with improved potential for preservation. The second is easier to find based on orbit-based morphology, such as large scale, fluvial depositional features, which may have formed even after life was present and offers, within a short driving distance, a wider window into the past of the planet if layered sediments can be accessed. A consideration related to morphology is the stability of the environment when life would have been present on Mars. Littoral environments can be quite unstable due to wave action and may therefore not contain a great concentration of traces of life and only limited quality for preservation. This limits the quality of the immediate surrounding of river channels both as habitat as well as a site for preservation. Target sites with the best chances of accumulating and preserving traces of life are therefore fluvial deposits such as deltas or fans. The inland delta of the Okavango river in Botswana may serve as an image illustrating such a combination of water, sedimentation, and teeming life on a young Mars. The wide area covered by a delta or fan with layers of sediment also increases the likelihood of a landing site offering a range of environments from the past, rather than just one, like a river. This, in turn, means less driving is required and one study site may show many depositional layers. The final factor to be considered when looking for a landing site is the history of the landforms since the deposition of the sediment. Even if young, rocks bearing traces of life are 2.5–3 billion years old. During this time, they would have ideally been buried mostly at a depth of several meters or more to protect them from destruction by radiation. To reach them now, even with the corers of NASA's Insight or ESA's ExoMars, the layers should not be deeper in the ground than 2 m. This requires current or recent erosion at the study site, moving host rocks closer to the surface.

Selecting landing sites for the search for traces of life at the surface or near subsurface environment emphasizes the need for understanding the landforms and stratigraphy of candidate sites. This, in turn, requires a reliable quantitative understanding of surface processes under reduced gravity.

BIBLIOGRAPHY

DiGregorio, B., 2010. The Microbes of Mars, Kindle edition published by author. Middleport, New York.

Geiger, H., 2009. Astrobiologie. UTB, Stuttgart, p. 235.

Hecht, M.H., 2002. Metastability of liquid water on Mars. Icarus 156, 14.

Jaumann, R., Köhler, U., 2013. Der Mars – ein Planet voller Rätsel. Edition, Fackelträger Cologne, p. 289.

Malin, M.C., Edgett, K.S., 2003. Evidence for persistent flow and aqueous sedimentation on early Mars. Science 302, 4.

Squyres, S., 2005. Roving Mars. Spirit, Opportunity, and the Exploration of the Red Planet. Hyperion, New York, p. 432.

Taylor, F.W., 2010. The Scientific Exploration of Mars. Cambridge University Press, Cambridge, p. 348.

Vago, J., Gardini, B., Kminek, G., Baglioni, P., Gianfiglio, G., Santovincenzo, A., Bayón, S., and van Winendael, M. 2006. Exo Mars. Searching for Life on the Red Planet, Report, pp. 16–23.

INTERNET RESOURCES

http://exploration.esa.int/mars/

http://insight.jpl.nasa.gov/home.cfm

http://photojournal.jpl.nasa.gov

http://planetologia.elte.hu

http://themis.asu.edu/node/5387

CHAPTER 4

Modeling Sedimentation

ABSTRACT

A key difference for the processes shaping the Martian surface driven by running water is gravity because both the flow hydraulics and the behavior of the sediment in water are strongly influenced by it. The relevant processes all act on Earth and numerical models have been developed to describe them. Many of these models are strongly empirical, but the way they are formulated implies that they deliver appropriate results for a range of gravities. Comparing the general effects of different gravities using the empirical models developed for Earth may serve as a first approximation, but requires testing for the much more specific environmental conditions distinguishing sedimentary environments likely to contain and preserve traces of life from those less promising for study by landers and rovers. Identifying the sensitivity of the output of common models to reduced gravity is, next to an introduction to conducting experiments, the main theme of this book. In this chapter, sediment settling and some common ways to calculate the settling velocity of sediment particles are presented to lay a foundation for the further experimental analysis on their applicability on Mars.

4.1 PARTICLE SETTLING

Interpreting a facies of sediments such as those of the Vingerklip shown in Figure 4.1 with regards to the hydraulic, and thus, implicitly the environmental conditions at the time of deposition, requires an

Experiments in Reduced Gravity: Sediment Settling on Mars. DOI: 10.1016/B978-0-12-799965-4.00004-2

Fig. 4.1. Vingerklip near Outjo, Namibia. This towering landform is the remnant of an eroded fluvial terrace. The deposits, formed by the river Ugap, are late Mesozoic and early Quaternary. Deposition was followed by erosion and formation of terraces during several cycles of changing sea and thus, base-level lowering of the Ugap. The outcrops offer a glimpse into the history and environment during the deposition of the rocks. In particular, the size can be used to assess the maximum kinetic energy for transport (just below the size of the largest boulders of a given layer) and the nature of the flow from which the deposition occurred: well-sorted indicates a steady flow of declining transport capacity, with continual deposition of the size class too large to be moved; poor sorting indicates a rather sudden loss of transport capacity and fairly shallow flows, limiting the sorting of sediment according to size. Credit: N.J. Kuhn.

understanding of the behavior of the particles in water. In case of the Martian conglomerates discovered by Curiosity, the particles range from fine sand to pebbles. The study presented in this book focuses on the settling of sediment from a water column. This process has a central role in sedimentology because it determines the size distribution of the particles forming the sediment. The close relationship of sediment particle size to the properties of the water body it settles from enables a reconstruction of flow properties such as velocity and depth. These, in turn, can be and have been used to assess the past environmental conditions on Mars, for example, by comparing Martian sediment to terrestrial analogs. Understanding sedimentation on Mars also enables the selection of landing and target sites for rock analysis with the greatest likelihood of containing traces of life. Testing the quality of settling velocity models developed for Earth when using them on Mars is therefore central for the current exploration of Mars. Studying sediment settling also enables an assessment of the quality of other models when applied on Mars because the interaction of sediment moving through water resembles conceptually runoff over a rough streambed or the entrainment of sediment particles by the shear stress of flowing water.

Particle settling velocity determines the time a particle requires to move through a layer of water from a given height to the base of the layer (see also Figure 1.8 in Chapter 1). If the water is moving in a lateral direction, the settling velocity also determines the transport distance of the particle traveling in the water. On Mars, information on particle settling velocity is required to calculate the movement of loose particles in running water such as the alluvial fans in Gale crater. The information can be used to assess environmental conditions, e.g., flow velocities, rates, and frequencies of runoff. Settling velocities are determined by gravity, size, shape, and density of the particle, as well as density and dynamic viscosity of the gas or liquid the particle is moving in.

4.2 MODELING TERMINAL VELOCITY OF SETTLING PARTICLES

The interaction of an irregularly shaped solid particle with water, especially when flow is turbulent, is complex. In theory, the movement of a particle in water can be described mathematically using Navier–Stokes equations. However, despite great increases in computational fluid dynamics (CFD) modeling, CFD has an extremely challenging time with irregularly shaped particles and predictions can deviate from actually observed values by 100%. Even in engineering, fluid dynamics around fairly simple shapes, unlike a sediment particle, are optimized by comparing model results with wind tunnel or flumes tests or plain test flights or drives. To achieve good results, a high spatial resolution of the CFD model is required, which needs a lot of computing power to solve. More important for sediment, measuring the shape of a particle and the distribution of electrostatic forces over its surface on a micrometer scale is difficult or not feasible, at least not for a large number of particles. Therefore, even for modeling the simplest case, a particle settling in fairly still water, simpler, calibrated models have been developed and the results are used as an index to describe particle behavior in water. One of the properties of particles in water used to describe the deposition environment is their terminal settling velocity. Apart from sedimentology, particle settling is also relevant in other fields of environmental sciences and management, for example, transport of solid particles between reservoirs geochemical cycles, the movement of particle-associated pollutants through the environment or simply the behavior of solids in sewage systems and waste water treatment plants. Developing working quantitative models for particle settling is therefore an important and still evolving theme in the geosciences; although first solutions were offered during the nineteenth century, continuous additions to the body of literature are still being made to the present day. A range of practical, empirically based solutions, based on simple versions of the Navier–Stokes equations and set boundary conditions exist.

The terminal settling velocity of a particle refers to the velocity achieved when forces acting on the particle are balanced (Figure 4.2). Two forces acting against each other can be distinguished:

F_g The force of gravity pulling the particle "down" and
F_d The frictional force, known as drag, acting against the acceleration by gravity.

F_g is determined by the gravitational acceleration acting on the particle, the density difference between particle and water and the volume of the particle. It can be written as

$$F_g = (\rho_p - \rho_f)g\ 4/3\ \pi r^3 \qquad \text{Equation 4.1}$$

with

F_g: gravitational force
ρ_p: density of particle

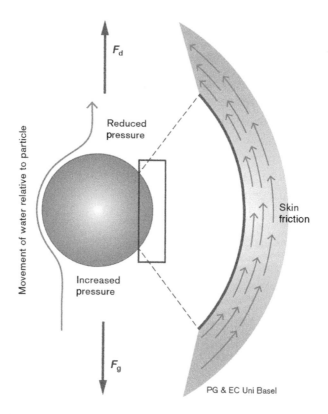

Fig. 4.2. Particle movement through water and associated forces: skin roughness affects the resistance of water passing by the surface of a particle and determines the velocity gradient of the passing water outward from the surface of the particle. The size and shape of the particle determine the pressure difference between front and back of a particle settling at a given velocity.

ρ_f: density of fluid in g/cm^3
r: radius of the particle in m
g: gravity in m/s^2

The drag or frictional force for the simplest case can be described by

$$F_d = 6\,\pi\mu r v_p \qquad\qquad \text{Equation 4.2}$$

F_d: frictional force
μ: dynamic viscosity kg/ms
r: radius of the particle in m
v_p: velocity of the particle

The equations show that for sediment in water F_g is constant and determined by the size and density of a particle. F_d, on the other hand, increases with settling velocity. The drag acting on a settling particle is also determined by two factors: first, the kinematics between the molecules and the sediment particle while moving around the particle, and second, the skin friction, i.e., the drag associated with the adhesion of water molecules to the sediment. The viscosity of the water, which describes the resistance of a liquid to deform, affects both types of friction and dominates the drag on the particle when the particle is small and does not generate a turbulent flow, a condition referred to as Stokes' flow.

The two equations can be combined into

$$v_s = \frac{2(\rho_p - \rho_f)}{9\mu} gR^2 \qquad \text{Equation 4.3}$$

with v_s being the terminal settling velocity.

For mineral grains settling in water the equation can be simplified to

$$w = \frac{RgD^2}{C_1 v_w} \qquad \text{Equation 4.4}$$

w: equation
R: difference between particle and fluid density
D: diameter
C_1: friction factor
v_w: kinematic viscosity of water

with w being the terminal settling velocity, R the difference between particle and fluid density, D the diameter, C_1 a friction factor with the value of 18 for a sphere, and v the kinematic viscosity of water. This equation is valid for single, small (>0.07 mm) spherical particles consisting of a homogeneous material with a smooth surface in laminar flow.

Larger and faster settling particles generate nonlaminar with large pressure differences between the front (increased pressure) and back (reduced pressure) of the particle (Figure 4.3). This state of particle–liquid interaction is called a Newtonian flow regime and exhibits a different

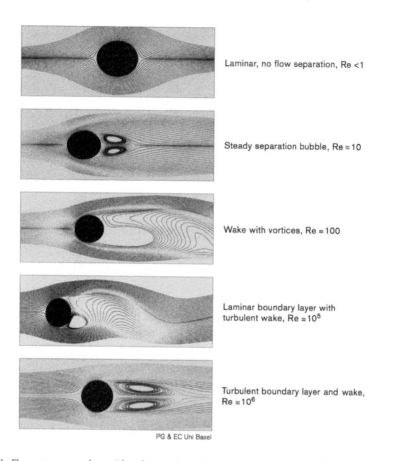

Laminar, no flow separation, Re <1

Steady separation bubble, Re ≈10

Wake with vortices, Re ≈100

Laminar boundary layer with turbulent wake, Re ≈10^5

Turbulent boundary layer and wake, Re ≈10^6

PG & EC Uni Basel

Fig. 4.3. Flow patterns around a particle with increasing settling velocity and corresponding Reynolds numbers.

relationship between forces induced by gravity and settling velocity. Stokes' law as stated above is not valid for Newtonian flow. For large sediment particles, i.e., >2 mm, the effect of turbulence on drag is much stronger than the viscous forces. Settling velocity can be estimated by

$$w = \sqrt{\frac{4RgD}{3C_2}}$$

Equation 4.5

w: settling velocity
C_2: spherical particles
R: difference between particle and fluid density
C_2 assumes a value of 0.4 for spherical particles, but can be increased to accommodate the effect of shape on drag.

Flow conditions around a particle can be estimated by the particle Reynolds number:

$$\mathrm{Re_p} = \frac{U_p D}{v}$$ Equation 4.6

$\mathrm{Re_p}$: particle Reynolds number
U_p: velocity of the relative particle–fluid
v: viscosity
D: diameter

with D being the particle diameter, U_p the velocity of the particle relative to the fluid, and v the kinematic viscosity of the fluid. Reynolds numbers below 100 can be considered laminar, above 100 turbulent, although a transition from 10 to 1000 is more accurate, especially for irregularly shaped particles.

Depending on the flow conditions around the particle, drag changes. The drag exerted on such a particle is usually not expressed as drag force F_d, but as drag coefficient C_d, enabling the description of drag as a particle property rather than a condition of the flow around the particle. The drag coefficient can be calculated by

$$C_d = \frac{F_d}{0.5\rho v^2 A}$$ Equation 4.7

C_d: drag coefficient
F_d: frictional force
A: reference area of the particle, i.e., the area projected into the main flow
ρ: density
v: viscosity

For a sphere, A would be the area of a circle corresponding to the radius of the sphere. Somewhat nonintuitively, drag C_d declines with increasing drag because it can roughly be seen as the proportion of settling velocity that would be achieved without friction. For a given particle, the drag coefficient is determined by flow velocity. As a consequence, particle Reynolds number and drag coefficient are also related to each other (Figure 4.4).

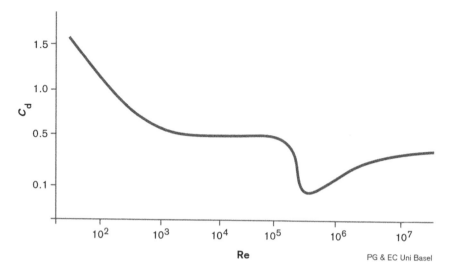

Fig. 4.4. Schematic representation of the relationship between particle Reynolds number and drag coefficient for a smooth sphere.

On Earth, one point along this relationship can be identified for a particle of a given size settling at terminal velocity when gravitational and drag forces are balanced. On Mars, this point would have to move toward lower Reynolds numbers because settling velocity is lower.

The nonlinear relationship between Reynolds number and drag coefficient illustrates the different nature of Stokes and Newtonian flow around a settling particle and the relative importance of viscous versus turbulent drag forces. To accommodate the change of flow regimes, a set of empirical equations has been proposed to estimate C_d for different Reynolds numbers (Appendix II equations).

The equations above show that drag increases with particle size. However, the effect is not linearly related to settling velocity or flow conditions, which has to be considered when using models for settling velocity and the way they influence flow kinematics. Using a sharp boundary for the application of either of the two models generates a zone of particle sizes in the fine-to-coarse sand range where both viscosity and turbulence act. Depending on the drag described in a model, different results for the settling velocity of the same-sized particle are the result (Figure 4.5). Therefore, a model for a wider range of particle

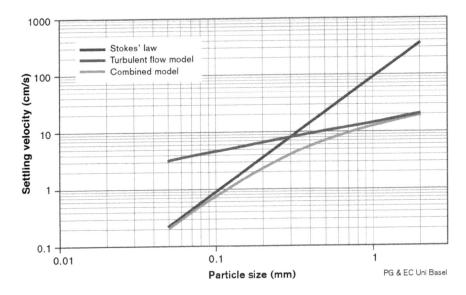

Fig. 4.5. Estimated settling velocities using Stokes' law, settling with turbulent flow, and a model combining both the conditions.

sizes can be used. An easy to use and well-calibrated version of such a model has been developed by Robert Ferguson and Michael Church and presented in the paper *A simple universal equation for grain settling velocity* in 2004. From here on called *combined model* balances the relevance of viscous versus turbulent forces by combining the two previous models into

$$w = \frac{RgD^2}{C_1 v + \sqrt{0.75 C_2 \, RgD^3}}$$
Equation 4.8

For small particles, the effect of viscosity on drag, described by the left term in the denominator is much larger than the one on the right. With D increasing by the third power, particle size dominates the settling velocity for larger particles. The use of this model has been tested by comparing results to a wide range of sediment settling velocities in the literature (Figure 4.5). This also enabled the fitting of C_1 and C_2 to actual sediment particles, i.e., deviating in shape from spheres. Fitting calculations to observations generally leads to higher values for both coefficients. This is plausible because both factors causing drag increase when particles exhibit a more complex shape.

4.3 SEDIMENT SHAPE AND CONCENTRATION

Developing different types of shape factors to describe the form of the sediment particles has accommodated the effect of irregular particle shapes. A factor often used is the Corey shape factor

$$Co = \frac{l_c}{(l_a \times l_b)^{0.5}}$$ Equation 4.9

with l_a, l_b, and l_c representing the longest, intermediate, and shortest axis across a particle.

Integrating the Corey shape factor into estimates for the drag coefficient leads to relationships such as

$$C_d = \frac{24}{Re} + \frac{0.5}{Co^2}$$ Equation 4.10

for Reynolds numbers <200 and small gravels (2–10 mm).

Figure 4.6 shows the relationship between the Re and C_d using the Corey shape factor. While it is clear that some improvement is made, the

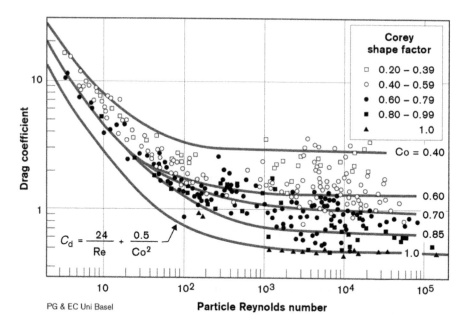

Fig. 4.6. Relationship between particle Reynolds number and drag coefficient for differently shaped sediment grains. Based on Julien (2010), p. 95. Based on Schultz et al. (1954).

uncertainties of describing the complex flow around irregularly shaped sediment particles remain obvious. Overall, this illustrates that practical models to describe settling of particles are still limited and rely heavily on calibration for a given scenario. The main reason for this is the difficulty of describing drag acting on a particle.

A final complication for estimating drag on settling particles is the interaction of the particles with themselves. This interaction can happen in two ways. First, high concentrations of fine, suspended particles may increase the density of the fluid and thus reduce the gravitational force acting on the larger settling particles. Suspended particles as well as dissolved substances may also change the viscosity of the fluid. Second, in a cloud of settling particles, the wakes generated by those ahead affect the drag on those behind. In theory, those following in the low-pressure zone should move faster than the settling velocity models suggest. As for the free settling of individual particles, semiempirical solutions correcting the drag coefficient over time have been developed to accommodate the change in gravitational forces and both viscous and turbulent drag.

4.4 IMPLICATIONS OF REDUCED GRAVITY FOR SEDIMENT SETTLING VELOCITY ON MARS

The empirical models for particle settling presented above have been used to assess sedimentation on Mars. Their applicability to Mars and other planets has not been questioned because they include all relevant factors describing the properties of sediment and water. However, the interaction between them is fixed for Earth, for example, the relationship between Reynolds number and drag. Applying these models on Mars requires either to assume that the implicit relationships between the factors controlling sedimentation on Earth and Mars are the same, or to recalibrate the models to avoid an undue extrapolation. Clearly, for major differences between planets, empirical models provide a first approximation. However, research aiming at finding traces of life and understanding Martian taphonomy has to work much more precisely. For example, the generally lower settling velocities on Mars (Figure 5.4) indicate that sorting of sediment from a given layer of water is less pronounced on Mars than on Earth. To be able to distinguish sediment deposited in a less habitable alluvial fan by episodic flows, interrupted by long dry periods, from more habitable riverbanks exposed to permanent or at least seasonal flows or standing water is of critical importance for landing site selection and rover or lander activities on Mars. Identifying the origin of sediments is therefore of critical importance for the current research on past Martian environments. In the light of sometimes small differences in sediment properties in environments of different habitability on Earth, ignoring gravity-induced differences in sedimentation between Earth and Mars carries the risk of missing the most likely sediment archive containing traces of life. The quality of the information available from Mars enables such detailed study. Therefore, testing and developing empirical models on sedimentation using experiments is of critical importance to achieve the aims of the current research strategies of Mars exploration.

BIBLIOGRAPHY

Dietrich, W.E., 1982. Settling velocity of natural particles. Water Resources Res. 18, 6.

Ferguson, R.I., Church, M., 2004. A simple universal equation for grain settling velocity. J. Sedimentary Res. 74, 5.

Hassanzadeh, Y., 2012. Hydraulics of Sediment Transport, Hydrodynamics – Theory and Model. In: Zheng J.-H. (Ed.), ISBN: 978-953-51-0130-7, InTech, Available from: http://www.intechopen.com/books/hydrodynamics-theory-and-model/hydraulics-of-sediment-transport.

Jerolmack, D.J., 2013. Pebbles on Mars. Science 340, 2.

Jiménez, J., Madsen, O., 2003. A simple formula to estimate settling velocity of natural sediments. J. Waterway Port Coastal Ocean Eng. 129, 70–78.

Julien, P.Y., 2010. Erosion and Sedimentation, Second Edition. Cambridge University Press, Cambridge, p. 371.

Knighton, D., 2014. Fluvial Forms and Processes: A New Perspective, Second Edition. Routledge, New York, p. 383.

INTERNET RESOURCES

http://www.calculatoredge.com

http://www.wired.com/2011/07/what-it-feels-like-for-a-sperm-or-how-to-get-around-when-you-are-really-really-small/

http://physics.info/drag/

http://www.rpi.edu/dept/chem-eng/Biotech-Environ/SEDIMENT/sedsettle.html

http://hinderedsettling.com/2013/08/09/grain-settling-python/

CHAPTER 5

Experiments on Martian Surface Properties and Processes

ABSTRACT

Experiments have a long tradition in geosciences and are considered to be a suitable tool to explore scenarios that are difficult or impossible to visit for an onsite measurement. However, experiments have limitations and results are often overinterpreted, or, as a consequence, received with undue skepticism. Therefore, some theory on use and limitation of experiments are required before moving on to designing instruments to test the effect of reduced gravity on sediment settling on Mars. This chapter presents first some general considerations researchers should take when planning an experiment. Based on these conceptual thoughts, the more specific aims of experiments in reduced gravity are defined, followed by a discussion of the constraints set for putting hardware onboard a reduced gravity flight.

5.1 EXPERIMENTS IN GEOSCIENCES

In Geomorphology, experiments have a long tradition because of the need to reduce complexity and simulate otherwise unattainable conditions. The great distance to Mars offers itself to use experiments to study the environment, calibrate models, prepare for robotic missions to Mars, or to support the interpretation of the data received from landers and orbiters. Strictu sensu, experiments involve controlled procedures to be carried out aiming at testing a hypothesis. This short review will focus on experiments in the

Experiments in Reduced Gravity: Sediment Settling on Mars. DOI: 10.1016/B978-0-12-799965-4.00005-4

classic, narrow sense, i.e., involving some sort of analog model or physical test in the real environment, but no pure numerical modeling. In environmental sciences, the term experiment is often used in a somewhat wider context, but three broad aims of "experiments" can be distinguished:

1. Actual experiments in the sense of the definition above, aimed at testing a hypothesis on the interaction of system components;
2. Process rate measurements in the field to quantify a conceptual model or a particular process in a given, qualitatively understood, landscape system; and
3. Measurements of an integrated response from a standardized setup to generate input data for quantitative modeling.

The major difference between these three aims lies in the knowledge about the components and their interaction on the studied system. A real experiment changes one component of the studied system, largely independent of scale, and identifies the effect on the output and evolving properties of the system. Components of a system include fluxes of energy or matter, either rate or quality, and a fixed property, such as shape or composition. The functioning of the studied system might be known to a detailed process level, but it may also be a gray or black box. Either an input flux into such a system is changed and the effects studied, or more often, a property controlling (supposedly) the functioning of the system is altered. This way, a hypothesis on the relevance of changing one component of the system for its overall behavior can be tested. Figure 5.1 illustrates this type of experiment in erosion studies testing the relevance of flow depth for erosion. An example for Mars are the tests conducted on the long-term effects of dust interaction in Martian storms on the suspended particles at the University of Aarhus. Here, a simple experiment rolling dust in glass vials for several months revealed the formation of larger particles as a consequence of repeated collisions. Since dust storms of Martian magnitude and duration do not occur on Earth, the extension of the time factor to explore the consequences of what happens in an otherwise simple and known system represents an experiment in its actual sense.

The second type of experiment is carried out when at least a conceptual understanding of the system behavior exists. The experiment is often required to measure a particular process rate to identify its relevance among other processes or its spatial and/or temporal pattern. In this case, the interaction between the components of a system has been

Fig. 5.1. Measuring the effect of flow depth on erosion in the Planetary Surface Process Lab of the University of Basel. The precisely controlled rainfall intensity, drop size, and kinetic energy are applied to a water level of increasing depth. Credit: B. Kuhn.

identified and the "experiment" focuses on collecting information on the rate of one or a set of different processes. Experiments on the erosion by sapping groundwater represent this type of experiment (Figure 5.2). Unlike the first type, they have to be downscaled and cannot simulate the Martian environment. However, they explore a realm which is not common on Earth, but plausible and in principle understood for Mars. Experiments on Mars that fall into this category are those carried out on the stability of water on the Martian surface. The general conditions under which water assumes a particular phase are known; however, the particular interactions on Mars have not been observed because they hardly exist on Earth. Therefore, experiments on melting ice or snow under the temperature, atmospheric pressure, and insolation conditions on Mars are required to test this particular scenario. This type of experiment could also be called a scientific simulation (Figure 5.3).

Fig. 5.2. Experiments on valley formation by groundwater sapping out of valley side slopes on Mars at the University of Utrecht. Experiments such as this one help to study the formation of landforms and associated surface processes on Mars that are widespread on Mars, but of limited occurrence on Earth. Credit: Wouter Marra, University of Utrecht, The Netherlands.

Fig. 5.3. Testing a Mars Rover. Credit: NASA, source; http://marsrover.nasa.gov/spotlight/opportunity/ b25_20040610.html

The final type of experiment can be considered a measurement of a predefined parameter value from a predefined system. The most prominent of these types of experiments are those used to collect data that feed into a database or model. Often the need for standardization or the presence of an existing data set leads to the design of such tests to ensure compatibility with existing information. Similarly, such tests can also involve the assessment of an error when using an established model outside the environment it is designed for. Experiments on Earth in the context of developing rovers for Martian exploration fall into this category because they largely seek to explore the limits of the engineering or the design criteria that have to be met to have a successful mission. Going to Mars represents an environment often outside common engineering design criteria and/or new technology needs to be used. This requires testing, but the complexity of the systems involved dictate the empirical context of the experiments. One could also call this type of experiment a simulation, but in an engineering context. As a consequence, the often-stochastic nature of these relationships necessitates constant refinement and collection of data for regions the model has not been used thus far.

The results generated by an experiment can often be used for more than one aim, i.e., their use and interpretation is associated with more than one of these categories of experiments. The experiments on particle settling in reduced gravity can fall in each of the three categories, depending on how the tests are designed and what is done with the data. Calibration of an existing model for terminal particle settling velocity falls clearly into the third category. Analyzing the differences between performances of a noncalibrated and a calibrated empirical model and trying to explain the differences in the context of Martian gravity fall into the first category. Using results to set boundary conditions of a computational fluid dynamics model based on Navier–Stokes equations would be in the second category.

5.2 DETERMINING THE AIM OF AN EXPERIMENT

Experiments always move in a triangle between realism, precision, and generality. To achieve the aims of a research project, it is important to distinguish between the different types of experiments and where they are situated in this triangle. It is therefore essential to define an objective for an experiment that is actually technically feasible. The information collected by achieving this objective should then contribute to generate an answer to the research question. The famous "42" delivered by the supercomputer Deep Thought after 7.5 million years of calculations to a bewildered crowd in Douglas Adams Hitchhiker's Guide to the Galaxy could be seen as such a poorly designed experiment. Three questions have to be asked to avoid such disappointment:

1. What should be measured?
2. What can be measured?
3. What was measured?

In an ideal world, we would have the instruments and resources to ensure that the answer to all three questions is the same. However, in most cases, this is not true. This can be illustrated with a simple environmental variable such as rainfall. Say the aim of the measurement is to know how much water was transferred from the atmosphere to the hydrosphere during a rainfall event over a known area. Measuring rainfall can be done in two ways: either by rain gauge, ideally connected to a logger for high temporal resolution during the event, or by remote sensing, either ground, air or earth orbit based. Rain gauges measure the amount of rainfall that falls into a funnel at a given point on the area it rains on, between two observations, or a series of recordings. This amount may or may not reflect the amount of precipitation on the area covered by the cloud that produced the rain. Errors could occur, for example, if wind drifts raindrops around the gauge, or when different intensities from the cloud introduce spatial variability that was not captured by the gauge and requires consideration when extrapolating from a point measurement to a larger spatial scale. Remote sensing covers the entire area the rain is falling on, but the detected value represents the reflection generated by the water drops at a certain height above the ground. The behavior of these drops between the time when they reflect a signal and when they would reach the ground is unknown. In both cases, with much calibration and the combination of both types of measurements, a model relationship that is assumed to give a satisfying answer to the question, how much rainfall has fallen over a specific area during a specific time, can be achieved. However, the answers to questions 2 and 3 should always be considered when using this result.

5.3 SIMULATING SETTLING VELOCITY ON MARS

Closing this section on experiments, the options for designing an experiment to simulate (for scientists and engineers) the effect of reduced gravity on Mars are examined against the backdrop of the questions raised above. The sediments examined by Curiosity and also those that from our current state of knowledge are most likely to bear traces of life, are sediments deposited by or from water. The most likely landing sites with such a watery past are those with some morphological evidence of fluvial activities because there, the features that can be recognized from orbit back the watery past. Clay-rich layers carry a greater risk of not having been deposited by water. The North–South dichotomy on Mars generated a lot of river channels that end in alluvial fans or deltas near the equator, which also offers the easiest (there are no easy ones) landing sites. This leaves sedimentary rocks such as sandstones and conglomerates, interbedded with some layers of fines that accumulated from shallow standing water, as likely rock types that will be examined for traces of life. To understand the formation of such rocks and the associated environmental conditions, settling velocities for Mars have to be known.

The aim of the research is therefore the calculation of the settling velocities on Mars that are realistic enough to support the selection of sites where rovers go to work after landing. The results should apply to actual sediment particles and not just involve theoretical considerations based on relative differences between calculations using terrestrial and Martian gravities.

Being unable to go to Mars directly, the use of available ways of calculating settling velocities (CFD modeling, semiempirical, or fully empirical model) have to be examined. CFD modeling would be preferable and is likely to deliver good results for regularly shaped particles. The main limitation is therefore the appropriate description of a real particle. Semiempirical models require information on drag coefficients for the combination of Martian settling velocities and particle size and shape. Neither is available, but could probably, at least for simple particles, be generated with CFDs to get at least a basic understanding of gravity effects on settling on Mars. The applicability of these results to more complex particles would require calibration and testing, as for any unstudied sediment on Earth in a particular hydraulic environment. A fully empirical model also necessitates calibration, but does not rely on any further relationships, such as those on Reynolds number and drag coefficient. The limitation, however, is that results are difficult to extrapolate. All three modeling approaches require measurements, to a different degree, to enable the generation of a reliable result. A practical question to ask at this stage is which way of calculating settling velocities on Mars involves the least effort, but the widest ranging result. Testing the use of a semiempirical model appears as such an easy first step because it can be done with the least number of measurements and generates a working model. Equation 4.8 presented in the previous chapter, repeated below for the ease of reading, represents the mathematical formulation of a model suited for this purpose, estimating settling velocity by

$$w = \frac{RgD^2}{C_1 v + \sqrt{0.75 C_2 \, RgD^3}} \qquad \text{Equation 4.8}$$

R: difference between particle and fluid density (t/m³)
g: gravity on Earth or Mars (m/s²)
D: diameter of particle (m/s²)
C_1: friction factor for viscous drag, 18 for spheres in water
v: kinematic viscosity of water (kg/m/s)
C_2: coefficient for turbulent drag, 0.4 for spherical particles

The equation represents an empirical model for the calculation of settling velocities on Mars (Figure 5.4). The quality of these calculations now requires testing. The main critical issue is the validity of the coefficients C_1 and C_2 for a different hydraulic environment. The equation also includes the key factors determining settling velocity: gravity, density of particle and liquid, as well as the particle size. This raises the question whether one of these factors that can be changed on Earth could be used to simulate sediment settling on Mars. The ratio of gravity between Mars and Earth is 0.38, so any other factor that could be technically reduced by the same proportion can be considered as a potential way to simulate settling under reduced gravity. With most relevant factors being both in the numerator and the denominator, any change has to have the same effect on the result than a change of

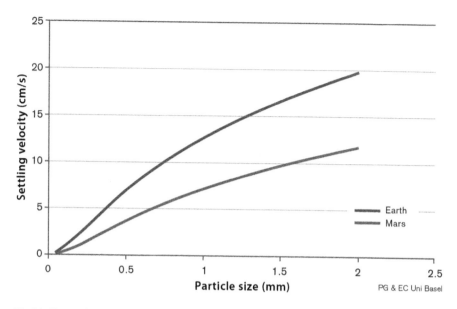

Fig. 5.4. Estimated settling velocities of sand-sized sediment for Earth and Mars using the combined settling velocity model.

gravity. Reducing particle diameter can therefore not be used to simulate Martian gravity because the combined effect in terms of the equation does not amount to the same result: a reduction of the squared diameter D to 38% in the numerator causes a decline by the square root of the cubed diameter in the denominator. This ratio is different from the effect of reducing gravity above and below the fraction line, except for one particle diameter when the two terms achieve the same value (Figure 5.5).

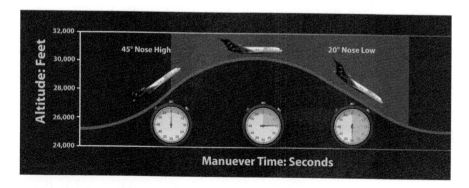

Fig. 5.5. Parabola flown to achieve weightlessness or reduced gravity for 20–30 s. Credit: The Zero-G Corporation.

A second option for simulating the effect of reduced gravity on particle settling could be achieved by reducing the density difference R between water and particle so that, the numerical solution of the settling velocity model generates the same result as a reduction of gravity. This approach does not generate a problem such as the reduction of size because the density difference is treated in the same way as gravity in the term describing gravitational force as well as the one describing turbulent drag. The density difference between sediment and water can be reduced by using lighter particles or heavier liquids. The density difference for most silicates is in the order of 1.65 g/cm³. Reducing it to 38% (= 0.63) would therefore require a liquid with twice the density of water or a particle with a density of approximately 1.65 g/cm³. Liquids of such a high density are hardly available. They also have a viscosity that differs from water, which inhibits their use for simulation of sediment settling on Mars. For sediment, light particles are available, e.g., a range of plastics that could even be shaped in a way to resemble sediment particles. The only problem for these are differences in cohesive forces that develop between the particle and the water. Water molecules are dipoles and some also separate into H^+ and OH. This leads to the formation of boundary layers of water molecules around sediment particles with charged surfaces. The thickness of this boundary layer depends on the electrostatic forces on the surface of the particle. On Mars, the potential for the formation of a boundary layer is exacerbated by the ions dissolved in runoff, assumed to be similar to water in playa lakes on Earth. Furthermore, because of the limited weathering, many sediment particles contain a lot of amorphous substances, i.e., clusters of molecules that have not formed into a crystal structure. Amorphous substances have a high proportion of surface areas with missing ions and therefore facilitate the formation of a boundary layer. Such boundary layers can also lead to a joining of particles by a process called coagulation, which effectively increases the particle size. For large sandy particles in the millimeter diameter range and formed by quartz, the effect of the boundary layer is negligible. However, below sizes of approximately a tenth of a millimeter, settling can be affected in a significant way. Therefore, the selection of a replacement substance with different interaction of surface and water should at least be tested before applying for simulations aimed at studying the effect of gravity on sediment settling and sorting.

The limitations of simulating sediment settling on Mars and on Earth lead to the question whether it would be possible to change g after all.

Increasing gravity is rather simply obtained by placing instruments in centrifuges. Reducing gravity, however, is more difficult and achieved in two ways. Moving to weightlessness and then, using a centrifuge enables testing any gravity below 1g. This procedure is actually used on the International Space Station. Choosing that option was (and still is), at least at the start of the research on the quality of an empirical model for calculating settling velocity on Mars, out of a realistic realm of costs. Short periods of weightlessness and reduced gravity can be achieved on Earth using drop towers and parabolic flights. Drop towers usually offer only a few seconds of weightlessness and achieving a reduced gravity involves complex braking of the dropped instrument. Parabolic flights generate up to 25 s of weightlessness or reduced gravity and are relatively easy to access. Exploring this option to measure particle settling under reduced gravity therefore appeared like a feasible way to move forward with our research.

5.4 DESIGNING AN EXPERIMENT FOR MEASURING SETTLING VELOCITY ONBOARD A REDUCED GRAVITY FLIGHT

Conducting surface process-related research during reduced gravity flights has been done before, most notably in recent years on the stability of loose material on Martian slopes by Maarten Kleinhans and his collaborators at the University of Utrecht. The key result of their work was the observation of a significantly greater static angle of repose of loose material on Mars, but a lower dynamic one for moving material compared to Earth. Such angles of repose were assumed to be independent of gravity until this experiment was done. A further analogy to our research is the empirical nature of the angle of repose. While models for their calculations exist, they are based on observations used to develop and calibrate empirical relationships. Similar to the discussion on the effects of reduced gravity on the terminal settling velocity of sediment particles in water in the previous section, their results indicate that models working on Earth should be used with caution on Mars. The implications of the effect of gravity on the angle of repose are not just theoretical, but may lead to a reassessment of the presence of water required for the formation of shallow landslides observed on Mars (Figure 5.6). This, in turn, may then lead to changing the interpretation of slides as indicators for climatic conditions on Mars at the time of their formation.

Fig. 5.6. Nephentes delta or other landslide: How much water was required to induce the landslides? Source: Mars Express Gallery. Credit: ESA/DLR/FU Berlin (G. Neukum).

The experiments on the angle of repose also illustrate that tests on surface processes onboard a plane flying 25-s parabolas are possible when the following three prerequisites are met:

1. The process must "happen" during this period of time.
2. An instrument containing the process and meeting the safety and space constraints has to be constructed.
3. It should be possible to observe the process and record its controlling factors during the flight.

Sediment particles achieve their terminal velocity in water within a few seconds and a short (<20 cm) traveling distance (Figure 5.7). Settling velocities are measured with instruments that range in length between 0.4 and several meters, depending on the quality of the measurement and the nature of the particles that are supposed to be measured (Figures 5.8 and 5.9). Sand and pebbles are better measured with longer tubes because their greater settling velocities require a longer distance for a good separation of size classes.

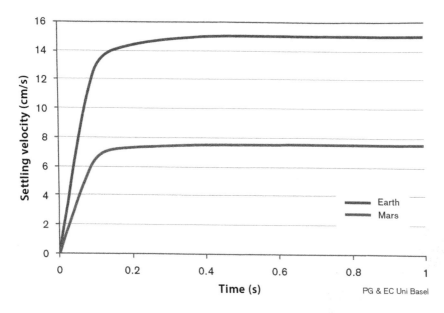

Fig. 5.7. Acceleration of a sphere with a diameter of 5 mm and a density of 2.65 g/cm³ settling in water on Earth and, assuming the same drag than on Earth, for Martian gravity.

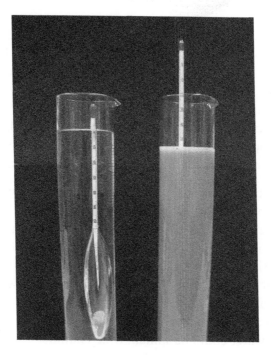

Fig. 5.8. Aerometers used to determine particle size or settling velocity based on the density change of the liquid. Note that the aerometer on the left is sunk into the clear water much deeper than the one on the right, which keeps sinking in with particles settling and declining density of the suspension. Pipetee measurements work on the same principle by taking a sampling at set time intervals at a given depth. The amount of sediment left suspended for sampling is assumed to consist of particles that are smaller than those that would have settled below this sampling point after mixing the suspension. Credit: Ruth Strunk, Uni Basel.

Fig. 5.9. Basel settling tube apparatus. Left: Sediment settling in the 1.8 m settling tube with injection device and collection tank. Right: Collection tank with sampling rotating sampling containers, set to collect sediment of a pre-set settling velocity. Credit: Yaxian Hu, Uni Basel.

Comparing existing techniques to measure settling velocity indicated that neither of these conditions posed a fundamental problem. The dimensions of an apparatus that can (affordably) be taken onboard a reduced gravity flight limit the length of the tube to 50 cm. Using the model for terminal settling velocities stated above, a first approximation yields settling times for sands on Mars ranging from 4 to 12 cm/s. A settling tube apparatus with 50 cm settling length and a settling time of 20 s are therefore sufficient to measure terminal velocities of sediment.

A key obstacle for the use of a settling tube is the way they are used to distinguish between different settling times. Small tubes such as the one seen in Figure 5.8, do not measure the distance a particle travels, but the change in the density of the fluid at a given depth. Based on the change of density per unit time, the size distribution of particles in the suspension can be calculated using a suitable equation, such as Stokes' law for fine sediment or one of the other models for larger material. This is done in open cylinders, which would not be possible onboard a plane, because water could be spilled, which hampers the measurement as well as posing a safety hazard. The technique used for larger settling tubes (Figure 5.9) has the same problem; these tubes are often open at the bottom, so that sediment can drop into collection containers that move along the opening in time intervals set to correspond to different settling velocities. We therefore had to develop a new approach to measure the movement of the particles in a closed instrument. Two approaches were developed and tested. The first one involved a portioning of the tube into several segments. Each of these chambers could be opened and closed separately, which enabled the release of the sample from the top with the lower chambers opened (Figure 6.3). The fractionation into different chambers was achieved by closing the other chambers after a given time. Dividing the 50 cm pipe into three equally sized receiving chambers and a sample chamber at the top would offer three 14-cm long settling tracks. Sand particles of 2 mm diameter would travel 40 cm in 4 s and reach the lower chamber, while small ones would have only moved 12 cm and thus, be caught in the upper or middle chamber. This timed multiple chamber method was tested and seemed appropriate to observe the general sorting behavior of sediment under different gravities.

The multiple chamber tube did not enable the measurement of terminal settling velocities of individual particles. To achieve this objective, recording their movement with small video cameras was tested. Provided the path of individual particles can be tracked in front of a ruler, the number of frames a particle requires to move a certain distance can be used to calculate the settling velocity. This method was also tested by filming particles settling in glass containers. The results were encouraging so that the construction of an instrument meeting all the criteria defined above appeared possible.

The process of designing an experiment ends with this general methodological test of "do-ability." For sedimentation on Mars it became clear that a comparative experiment can be carried out and help to understand deposition on Mars. This, in turn, will contribute to our search for environments with potentially life-harboring scenarios on Mars.

BIBLIOGRAPHY

Hecht, M.H., 2002. Metastability of liquid water on Mars. Icarus 156 (2), 373–386.

Marra, W.A., Braat, L., Baar, A.W., Kleinhans, M.G., 2014. Valley formation by groundwater seepage, pressurized groundwater outbursts and crater-lake overflow in flume experiments with implications for Mars. Icarus 232, 21.

Nørnberg, P., Bak, E., Finster, K., Gunnlaugsson, H.P., Iversen, J.J., Knak Jensen, S., Merrison, J.P., 2014. Aeolian comminution experiments revealing surprising sandball mineral aggregates. Aeolian Res. 13, 77–80.

Richter, G., 1981. Recent trends of experimental geomorphology in the field. Earth Surf. Process. Landforms 6, 215–219.

Schumm, S., Mosley, M.P., Weaver, W.E., 1987. Experimental Fluvial Geomorphology. Wiley-Interscience, p. 413.

Slattery, M.C., Bryan, R.B., 2006. Hydraulic conditions for rill incision under simulated rainfall: a laboratory experiment. Earth Surf. Process. Landforms 17, 19.

Slaymaker, O. (Ed.), 1991. Field Experiments and Measurement Programs in Geomorphology. Balkema, Rotterdam, p. 224.

INTERNET RESOURCES

http://marsrover.nasa.gov/spotlight/opportunity/b25_20040610.html

http://www.dlr.de/mars/en/DesktopDefault.aspx/tabid-388/7423_read-12328/gallery-1/gallery_read-Image.8.5367/

CHAPTER 6

MarsSedEx I: Instrument Development

ABSTRACT

A crucial step between gaining meaningful results from an experiment is the design of the hardware and the measurements that are conducted with it. This requires a clear understanding of the processes that are measured and their controlling factors, as well as the context of the measurements and their relationship to reality. This chapter therefore starts with a definition of the scientific aims of the MarsSedEx I mission, followed by a description of two settling tubes and the measurements that could be conducted with them. For both experiments, the quality of the delivered information with regard to the objectives of the measurement and thus the overarching aim of the mission are explained. The chapter closes with more practical questions related to putting a scientific instrument onboard a reduced gravity flight.

6.1 SCIENTIFIC AIMS AND DESIGN OF THE MarsSedEx I INSTRUMENTS

The analysis of particle settling on Mars and the numerical models that are available to assess it presented in the previous chapters demonstrate the need for measurements of settling velocity under reduced gravity and indicate that this should be possible during a reduced gravity flight (Figure 6.1). The MarsSedEx series of missions comparing terrestrial

Experiments in Reduced Gravity: Sediment Settling on Mars. DOI: 10.1016/B978-0-12-799965-4.00006-6

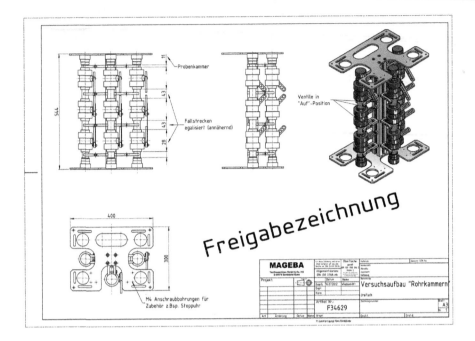

Fig. 6.1. The first engineering sketch of the MarsSedEx I instrument carrying frame and multiple chamber settling tube prototypes. Credit: Kai Wiedenhöft, Mageba GmbH, Germany.

and reduced gravity was launched to move from theory to reality. The series of flights is at this stage open ended, but follows a stepwise strategy to build technical expertise and accumulate data on sediment settling on Mars. The experiments conducted for MarsSedEx I aimed at:

1. determining whether the terminal settling velocity of glass spheres of 1000 μm diameter can be measured during a reduced gravity flight for lunar and Martian gravity (*One-Chamber Settling Tube (OCST) Test*); and
2. measuring whether the theoretical considerations on the unmixing of sediment particles during settling can be observed in reduced gravities (*Three-Chamber Settling Tube (TCST) Test*).

Particle settling velocity is commonly determined by measuring settling velocities in settling tubes. Conventional measurements aimed at determining particle size use tubes of 40–50 cm length and determine the change of fluid density or the size distribution of particles at a particular depth, both over time (for such a tube, see Figure 5.8).

The measurements last up to several hours, even a day, so that this approach is not feasible for a reduced gravity flight. Fractionation of sediment by settling velocity is also done by settling tubes, but requires the use of tubes up to 2-m long and also measurement intervals of minutes to hours (instrument presented in Figure 5.9). The main constraint on a reduced gravity flight is the short duration of the reduced gravity. Any measurement therefore has to be conducted within 20–25 s. A key objective of the MarsSedEx I experiments was therefore to determine whether a settling tube apparatus could be designed that would deliver a meaningful result within this short period of time. In addition, the relevance of the theoretical considerations on sediment settling derived from empirical models developed for Earth was tested. Two settling tubes were designed to address these objectives.

6.2 ONE-CHAMBER SETTLING TUBE (OCST) EXPERIMENT

The experiment aimed at testing the feasibility of terminal velocity measurements under reduced gravity. If successful, future tests could be designed to measure a range of particle settling velocities on Mars, to develop further instruments for related parameters, such as the parameters surface roughness, erosion, and transport, and to inform models of both the formation of sedimentary deposits as well as landform development, e.g., alluvial fans on Mars.

The major constraint during reduced gravity flights is the limited time available for a measurement. Furthermore, the instruments have to be closed for safety so that no liquid is lost. The substances involved in the experiments also have to conform to flight safety regulations. Conventional aerometer (Figure 5.8) or pipette measurements of particles size would work in reduced gravity in theory, but the development and flight cost of a pipette apparatus working in a closed system was prohibitive for an initial test. A simpler way for a feasibility test is the direct observation of settling particles using a small video camera to record their movement against the background of a ruler. After landing, the number of frames required for a particle to move a certain distance can be counted to determine the settling velocity. The only limitation of this approach is the resolution of the camera and thus the smallest

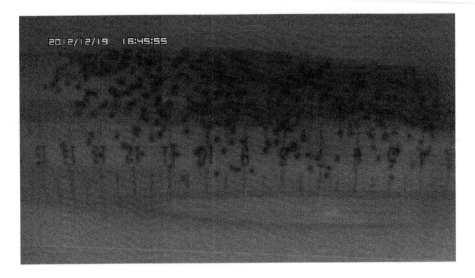

Fig. 6.2. A frame taken from a video clip showing a cloud of pink reference spheres settling in a test chamber.
Credit: B. Kuhn.

particle size that can be used. Comparing the resolution (on the order of 1 megabyte) of small action cams showed that close-up recording (10-cm distance) would enable, in theory, the distinction of particles on the order of 0.1 mm in diameter. Testing two cameras, a GoPro Hero and a HY001HD police cam showed that particles of 0.5-mm diameter and a strong color contrast to the background could be still identified (Figure 6.2).

Spherical particles of 0.5-mm diameter with the density of quartz achieve terminal velocities in water of $20\,°C$ of approximately 8, 4, and 2 cm/s on Earth, Mars, and Moon, respectively, using the combined model developed by Ferguson and Church (2004). They reach these velocities well below the 20 s settling time and within a settling distance of less than 20 cm. The latter is important for a reduced gravity flight because it determines the length of the settling tubes, which would ideally be below 50 cm to keep the flight cost low and development of a stable structure to carry them meeting the safety criteria simple. A further point to consider is that gravity takes a couple of seconds to drop and stabilize, so settling tests had to be designed to work in less than 20 s to accommodate these phases of transition. A second concern regarding the duration of settling is the effect of varying gravity during the reduced gravity parabolas.

Table 6.1. Properties of Particles Used for MarsSedEx I Tests

Sample	D_{50} (μm)	Density (g/cm³)	Origin
212–250 μm spheres	225	2.39	Microspheres–nanospheres Cat. No: 156015-50
355–425 μm spheres	388	2.47	Microspheres–nanospheres Cat. No: 156121-50
500–600 μm spheres	532	2.48	Microspheres–nanospheres Cat. No: 155125-50
710–850 μm spheres	767	2.47	Microspheres–nanospheres Cat. No: 155629-50
1000–1180 μm	1020	2.33	Microspheres–nanospheres Cat. No: 155833-50

Larger, fast settling particles are less affected than smaller, slower settling ones. Finally, the particles have to be detected on the video even under the possibly varying lightning conditions during the flight. With these restrictions in mind, pink glass spheres of 1-mm diameter were chosen for the MarsSedEx I tests (Table 6.1). On Earth, they have a terminal settling velocity of approximately 21 cm/s and because of their pink color were easily distinguishable on the video (Figure 6.3).

The OCST used for MarsSedEx I is made of Plexiglas and is 45-cm long with an inner diameter of 5 cm (Figure 6.3). At the top end, a plastic ball valve distributed by SIBO is clued to the tube. The inlet of the ball valve and the lower end of the tube are closed by a screw cap. The top cap fixes a plastic plate to an O-ring, creating a supply chamber for the sediment tested (Figure 6.4). The bottom one is custom made to fit onto the thread of the Plexiglas pipe. The tube itself is fixed to the mounting frame by aluminum clamps. The tube is filled with water. The space between the actual ball in the valve and the top screw cap serves as a supply chamber for the pink glass spheres. The load of the glass spheres covers the closing ball. This amount of "sediment" provides a spread-out cloud of particles, provided the ball valve is opened slowly, for recording by the video camera (Figure 29).

Originally, only one settling tube test was planned for the MarsSedEx I flight (ZG 318). However, due to an unexpected early termination of the first flight, reloading the settling tube, and a second test were possible a day after the first flight (ZG 319). This way, the terminal velocity both for lunar and Martian gravities could be measured. The procedures required to operate the OCST are described in Chapter 7, while Chapter 9 focuses on the results of the OCST experiment.

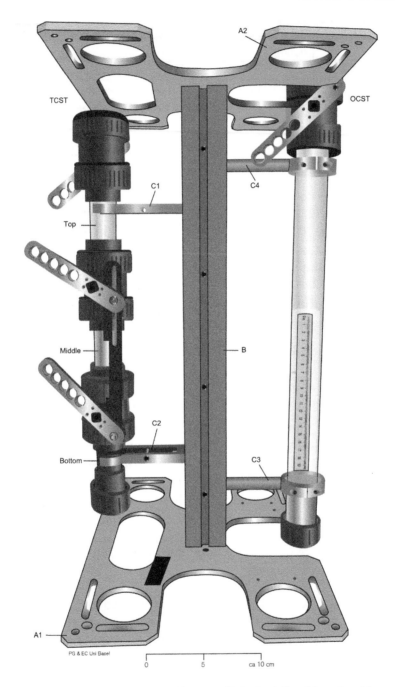

Fig. 6.3. The OCST and TCST settling tubes used for MarsSedEx I. A1 and A2: endplates; B: upright support beam; C1 to C4: holding clamps; TCST; top, middle, and bottom: sample chambers of TCST separated at 15 and 30 cm distance from supply chamber at the top; and OCST.

Fig. 6.4. Glass spheres loaded in the top chamber of the MarsSedEx I TCST. The same chamber was used for the OCST and for MarsSedEx II. On this image, the mixture of spheres used for the TCST is seen, for the OCST, only the large pink reference spheres were used. Credit: B. Kuhn, Uni Basel.

6.3 THREE-CHAMBER SETTLING TUBE (TCST)

The TCST consists of a closed plastic tube of 45-cm length and 5-cm diameter mounted in an upright position (Figures 6.1 and 6.3). The tube is split into four chambers separated by ball valves. The top chamber contains 2.5 g of glass spheres, divided into five classes of 0.5 g each and ranging in nominal sizes between 212 and 1180 μm (Table 6.1). At the beginning of the experiment, the top chamber is closed by the valve while the other three are open. Upon the onset of reduced gravity, the top chamber is opened for 15 s so that the glass spheres can settle through the water column. At the end of the settling period, the lower three chambers are closed. After the flight, the frame is taken from the airplane and the chambers are emptied. The size of the glass spheres in each chamber is determined. Based on a comparative experiment under terrestrial gravity, differences in the size distribution between the original mixture and the particles collected in each chamber are indicative of the effect of reduced gravity on sorting of sediment under reduced gravity. In addition, the experiment serves as a feasibility test for more complex low-gravity experiments simulating sedimentation processes on Mars. As for the OCST, two measurements were possible during the second MarsSedEx I flight. Switching the gravity of all tests from lunar to Martian and vice versa was not possible for operational reasons, so a second TCST test was planned for ZG 319. This second measurement under lunar conditions would deliver an indication of the replicability of the results achieved in reduced gravity.

6.4 STRUCTURAL STABILITY, SAFETY, AND FEASIBILITY CONSIDERATIONS

Research flights in reduced gravity require a particular set up of the instruments to conform with stability and safety regulations as well as the cost structure of the flight and the operators' ability to perform. In this section, a summary of the considerations for MarsSedEx I is given. For MarsSedEx II, the actual research proposal submitted to Zero-G with the required structural, safety, and operational information is added in Appendix III Research Proposal MarsSedEx II.

Both MarsSedEx I and II were designed to be operated by one operator-scientist and to stay within given limits of weight and size. The MarsSedEx I settling test instrument has the dimensions and weight offered by Zero-G (15″ length × 16″ width × 21.5″ height, 25 pounds, corresponding to 38, 41, and 55 cm, respectively and a weight of 11.3 kg) for a single-operator and fixed instrument research flight (Figure 6.5). The support structure of the apparatus consists of a center square beam and a bottom and top plate. The settling tubes are mounted each onto a side of the central square beam. Beam, plates, and clamps connecting tubes and beam are manufactured from aluminum; the tubes, caps and valves from plastic and Plexiglas. A certified engineer (Dipl.-Ing. Kai Wiedenhöft, Mageba Maschinenbau GmbH, Bernkastel-Kues, Germany) was responsible for the original design of the support frame and TCST to ensure that stability criteria were met. Both, MarsSedEx I and II support structures were manufactured by Mageba Maschinenbau GmbH, as was the first set of TCST settling tubes (Figure 6.1). Based on the documentation of these designs, the OCST and the structures required for the instrument installed in the periphery of the settling tubes were

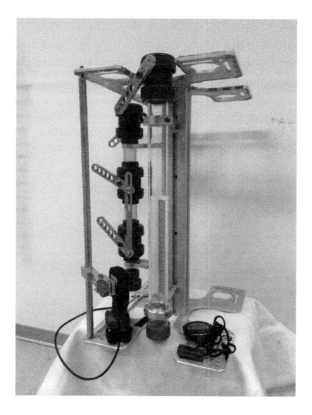

Fig. 6.5. The MarsSedEx I settling tubes mounted in the instrument frame with camera, g-loggers, and stopwatch. Credit: N.J. Kuhn.

designed and fabricated in the GEO work workshops of the University of Basel. The tubes are filled with water and sediment.

In addition to the frame and tube, an iPhone in flight mode, used as a g-meter with easily readable display, a g-logger (WSR-145), a stopwatch, and two cameras were mounted in the frame (Figure 6.5). The watch and g-meter are used as an aid for the operator, the g-logger to record the actual gravity during the flight at 1-s intervals. One camera is used to record the settling of particles while the other recorded the movements of the operator and an experiment not covered in this book. The measuring and recording instruments were switched on before the first parabola was flown. All electrical equipment (cameras, digital clock, g-logger, iPhone) used for this experiment was commercially available and had not been changed from its original design. It was therefore safe to carry and use on an airplane. No electrical power supply from the airplane was required. None of the materials used for this experiment poses a hazard according to the FAA guidelines and the Material Safety Data Sheets (MSDS) Bibliography used in the United States.

BIBLIOGRAPHY

Ferguson, R.I., Church, M., 2004. A simple universal equation for grain settling velocity. J. Sedimentary Res. 74, 5.

Grotzinger, J.P., Hayes, A.G., Lamb, M.P., McLennan, S.M., 2013. Sedimentary processes on Earth, Mars, Titan, and Venus. In: Mackwell, S.J. et al. (Ed.), Comparative Climatology of Terrestrial Planets. Univ. of Arizona, Tucson, pp. 439–472.

Hu, Y., Fister, W., Rüegg, H.-P., Kinnell, P.I.A., Kuhn, N.J., 2013. Section 1.1.1: The use of equivalent quartz size and settling tube apparatus to fractionate soil aggregates by settling velocity. In: Clarke, L.E. (Ed.), Geomorphological Techniques. London, UK, pp. 1–9.

Komar, P.D., 1979. Modes of sediment transport in channelized water flows with ramifications to the erosion of the Martian outflow channels. Icarus 42, 13.

Loch, R.J., 2001. Settling velocity – a new approach to assessing soil and sediment properties. Computer Electronics Agriculture 31, 12.

INTERNET RESOURCES

https://www.youtube.com/watch?v=ZLHRSFvK9Uw

http://www.gozerog.com/index.cfm?fuseaction=Research_Programs.welcome

CHAPTER 7

Preparing and Flying the MarsSedEx I Research Flight

ABSTRACT

While most experiments are, on purpose, in laboratories with a controlled, but forgiving environment, flying and operating an instrument onboard a reduced gravity flight is almost the opposite. Greatest possible accuracy of observations has to be achieved in a rather unusual environment, combined with the stress of a very limited number of replicates that can be conducted. Added to this is the need to travel with all the equipment to the airport where the flight takes place, including passing all security and safety checks. Overall, our experience tells us that experiments in reduced gravity combine the pressures of the laboratory with those doing fieldwork. Therefore, this section introduces the procedures involved in getting ready to fly and shares some experiences and advice on conducting experiments in reduced gravity. As for the more general thoughts on experiments in geosciences, they are also intended as a more general advice, especially to young researchers, going out and about exploring, either in the lab, the field, or both.

7.1 GETTING READY TO FLY

Building an instrument for a reduced gravity flight is one task, but getting it ready to fly involves more. In the following section, the activities and our experience is documented, mostly information that the common researcher would not put into a research paper. It is placed here to be of use not only for the few who take their research onboard a reduced gravity flight, but also to every researcher going out and about to places when everything has to work.

Experiments in Reduced Gravity: Sediment Settling on Mars. DOI: 10.1016/B978-0-12-799965-4.00007-8

Low/zero gravity research flights offered by Zero-G start with a Flight Readiness Review (FRR) on the day before the flight. The FRR involves a safety and feasibility inspection of the research equipment set up ready to use in a hangar at the airport where the flight is conducted.

During the FRR, not only the safety of the equipment itself, but also the activities of the researchers flying and using the equipment are discussed, including contingency plans in case the equipment fails. For the MarsSedEx I apparatus the major concerns involved how to deal with water leaking from the settling tubes. As a precautionary measure, a large towel and plastic bag to cover the apparatus were taken on-board for the flight. Further safety concerns included sharp edges of the top plate of the apparatus. These were covered with several layers of duct tape. To ensure passing the review and getting the intended experiments done, a set of spare parts and tools should be taken along since usually something that has never broken before will do so shortly before the flight. For newcomers to the flights, a person on the team being able to make quick drives to the next hardware store is strongly recommended (Figure 7.1)!

Fig. 7.1. Inspection of the MarsSedEx II instrument during the FRR (during the MarsSedEx I FRR, we were too nervous to take pictures…). Credit: Bryan Rapoza – Avenfoto for Zero-G.

The FRR also offers an opportunity to get the support of the flight crew. Apart from pilots, the crew consists of a flight director who is responsible for the activities in the cabin, at least one FAA-certified flight attendant and several coaches who support both flight director and researchers. Explaining your activities to the flight director and coaches during the FRR identifies critical points, especially to first-time participants, and enables the researchers to get the support of a coach, if required. For MarsSedEx I, for example, there was a chance that pushing the handles of the TCST at lunar gravity would lead to an upward movement from the operator. Therefore, a coach was assigned to push the operator onto the shoulders to keep him seated during reduced gravity (Figure 7.2). Talking to coaches and flight director, we also realized that we had to find a different way of measuring water temperature than planned. Originally, a laser thermometer was supposed to be aimed at the settling tube during the flight, assuming the temperature of the tube would correspond to the water inside. However, shining a laser around in the somewhat unstable and crowded cabin was considered unsafe. Besides, with a drop in temperatures during the flight, the outside of the tube might actually be colder than the water inside. A quickly designed alternative involved a sprint to a supermarket and a pet shop to purchase a children's drink bottle, appropriately with a Mickey Mouse

Fig. 7.2. Author supported by a coach during MarsSedEx I to ensure that turning the handles of the TCST did not end up in floating away and crashing down. Credit: Zero-G.

print, an aquarium thermometer, a large breakfast bag, and some duck tape. This was combined into a makeshift container to record the correct water temperature immediately before and after the measurements.

After inspection, the instrument is mounted in the plane. The time between mounting and flight can easily amount up to 20 h (Figure 7.3). Depending on the nature of the experiment and the instruments used, this requires consideration when designing and preparing the instrument as well as the activities conducted during the flight. In the case of Mars-SedEx I (and II), all electronic devices had to be powered down, apart from flight safety during takeoff and landing, to conserve the power available from the batteries. Further considerations for both MarsSedEx I and II involved checking the positions and settings of cameras after reaching cruising altitude because there is a chance that during the loading process somebody may have touched and displaced them. Powering all the instruments up and putting them back into the proper setting may also require some time between takeoff, reaching cruising altitude, and the beginning of the parabolas. The time required should be tested in advance and given some extra time to account for the unusual environment

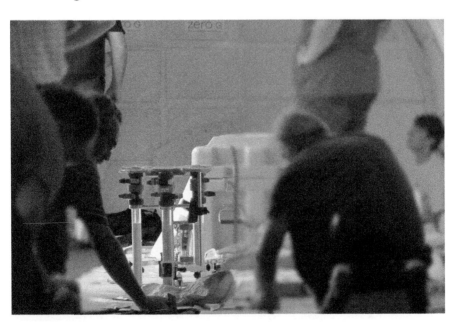

Fig. 7.3. Mounting the MarsSedEx II apparatus; the plane can be quite busy, cramped, and crowded, so both attention and patience are required. Credit: N.J. Kuhn.

even at regular cruise flight conditions. The duration of the required preparation time has to be communicated to the flight crew to ensure that everything is set up. It should also be noted that the plane is largely powered down before the flight, i.e., no air conditioning. Depending on the ambient temperature outside, this can lead to a notable drop or increase of water temperatures in settling tubes depending on the initial water temperature. More importantly, temperatures changed during the flight due to the lack of a heating system, they actually dropped during both MarsSedEx I and II. If experiments, such as ours, are sensitive to temperature, a constant control and logging is required and a cross-check with the temperatures used for the ground tests is required.

Security and safety procedures of a low-gravity flight correspond to those of a regular flight. This can be a problem for the equipment taken on-board on the day of the flight. In case of MarsSedEx I and II, the limitations of the inspection equipment available for security checks (just handheld metal detectors), as is often the case at regional airports without regular passenger flights, have to be met. Any equipment that seemed suspicious to the FAA safety officers was not allowed on-board, no arguing, even after bringing the material from Europe as cabin luggage. In our case, the water in the settling tubes had to be provided by Zero *g*. Spare sediment samples were not allowed through security for the second flight of MarsSedEx I and we had to rely on the spare sample already on-board. To avoid any problems, equipment should be inspected during the FRR and left on-board the day before the flight or cleared during the FRR for bringing on board during the day of the flight. After FRR and loading, not much is left to be done, so a final practice for the activities involved during the flight is suggested because the actual situation on-board the plane is now known. This should be balanced with getting some relaxation to be able to perform well during the flight.

7.2 MarsSedEx I FLIGHT ACTIVITIES

The flight day starts similar to more or less every other flight, checking in and a security check. After take-off and reaching cruising altitude some time remains to prepare the experiments. For MarsSedEx I and II between 5 and 10 min were required. We strongly suggest having a checklist to ensure that absolutely everything is set up correctly and that the efforts of many people and funding bodies do not fail just because the flying operator forgets to press the right button on a camera. Physically we recommend a check list that has few items on a page, pages that can be flipped easily and are bound in a way that enable fixing it to your body or the instrument in a way that makes it easy to see and read. In weightlessness, sheets float, so fixing them is important, but should not hinder turning them too much. A further point to consider is that the order or the way an operator conducts his or her activities during the flight might change as a consequence of the FRR or the way research instruments are mounted in the plane. It is therefore highly advisable to be able to rewrite or amend a flight plan after the FRR and mounting in a way that does not compromise its use during the flight. Figure 7.4 shows the revised flight plans for MarsSedEx I and the flight manual used during MarsSedEx II (Figure 10.5), illustrating the learning curve we went through. This point of the flight is also a good opportunity to do some last minute practicing of the actual moves you have to do. Get in your high *g* position and then move into the low/zero position, reach for the handles/buttons/tools you have to work. Try to find out whether these moves work as you have practiced them before and talk about them with the coaches. A good communication about what you want to do at this stage to get or adjust the support by the coaches increases the chances of success and limits the number of surprises.

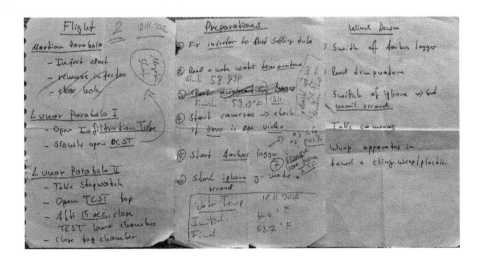

Fig. 7.4. Revised flight plans for MarsSedEx I.

The first parabola is always a mixture of excitement and nervousness, so it is important to focus on the set of actions that you have to do. Relax and repeat to yourself what activities you have to do, if possible add a short rehearsal. Due to the limited time at reduced/zero g, it is useful to memorize the activities for a given parabola. However, due to the unusual environment and the likely nervousness, it is useful to have a page in the flight plan for every parabola. It should be noted, however, that lying flat on your back is the best position to get through this high g part of the flight (see Chapter 8 on the human dimension of reduced gravity flights) and that reading about what to do next when reduced/zero gravity kicks in is difficult at $2g$. Therefore, the flight plan for the parabolas best consists of a cue card for each parabola that can be looked at quickly at any time of the parabola. Once reduced/zero gravity sets in, starting experiments should not be rushed so that the gravity experienced by the plane has stabilized.

The flight activities for MarsSedEx I are listed below to give an impression of the tasks that had to be carried out. The flight activities can be split into three parts: preparabola, parabola, and postparabola. Preparabola activities included:

1. Before lunar and Martian gravity parabolas: start cameras, clocks, digital recorder, gravity meter, and iPhone; record water temperature.
2. Ensure easy access to handles of all instruments.
3. Take stopwatch, set at zero, and hold in a way that does not hinder starting TCST.
4. Get into a comfortable position for $2g$ before first parabola.

Three parabolas offered measurements under reduced gravity during the November 18, 2012 flight (ZG 318), one with Martian and two with lunar gravities. To adapt to the new and peculiar conditions, no MarsSedEx-related tests were conducted during the first parabola. The activities during the following two parabolas consisted of:

1. First lunar parabola: open ball valve separating top chamber of TCST as soon as lunar gravity has been achieved (iPhone g-meter below $0.18g$). Start stopwatch at the same time. Close lower two ball valves of chamber 15 s later.
2. Second lunar gravity parabola: open ball valve of settling tube OCST.

The reduced/zero gravity parabolas during Zero-G research flights are flown in sets of five, interrupted by a break to prepare and adjust. The three lunar and Martian gravity parabolas on the November 18 flight were followed by two zero *g* parabolas. After those, the break was used to switch off the recording equipment and secure the storing devices. We suggest keeping the amount of equipment that is taken off the plane immediately after the flight as limited as possible in case a second flight is required. Everything that moves out has to pass a security check again, which may cause problems because of the constraints during the flight day described above.

INTERNET RESOURCES

http://www.gozerog.com/index.cfm?fuseaction=Research_Programs.welcome

The Human Dimension of Reduced Gravity

ABSTRACT

The world of scientific enquiry does usually not report the measures taken to ensure the wellbeing of its community. In most cases, life-threatening situations are avoided and, especially in the field, work is conducted to the physical and mental limit of those involved to get the most out of the precious time available. This approach, and the wide range of interpretations and adjustments it experiences by those conducting the field work, is not feasible for a reduced gravity flight because the opportunity to do measurements are few and costly. Therefore, this chapter on doing as much as possible to be fit for the task and considerate to fellow flyers has been added to the book. Apart from being a slightly lighter hearted break from the science, it may also serve as advice for those planning a first "big trip" to be prepared.

8.1 THE HUMAN DIMENSION

Before cracking on with the results of the MarsSedEx I flight, some more general advice on the experience of a reduced/zero g flight appears appropriate. As for the chapter on flight preparations, it is also intended to serve as a guide to research not quite on the planet anymore, but to hopefully contribute to the success of any mission a reader may venture on. Among astronauts and researchers, zero g planes are nicknamed the "Vomit Comet." Therefore, a few remarks on the more human dimensions of reduced/zero g flights seem helpful. For a full account see "Throwing up and down" in Mary Roach's book *Packing for Mars*. However, if you are anxious about how you will fare during your first zero g experience, I suggest reading this chapter of the book after the flight (Figure 8.1).

Motion sickness in its full range of varieties is common during zero g flights. The main reason is the confusion caused by the changing gravity

Experiments in Reduced Gravity: Sediment Settling on Mars. DOI: 10.1016/B978-0-12-799965-4.00008-X

Fig. 8.1. Author's feet floating in front of MarsSedEx I. Credit: N.J. Kuhn.

to the vestibular system and the information on the position of your body relative to the space surrounding you generated by eyes and brain. In many ways, it feels similar to seasickness, but shows also some distinct differences. On reduced/zero *g* flights, motion sickness usually sets in after 10–15 parabolas. Many participants are affected by it to various degrees. Even seasoned astronauts express both their dislike of zero *g* flights as well as the strong recommendation to take motion sickness pills. Based on our experience, ensuring that the research activities on board are not affected by motion sickness is the most important prerequisite to a successful flight. We therefore strongly advise every researcher participating in a zero *g* flight to take appropriate and adequate medication. While motion sickness pills have some side effects, for example, being tired, they outweigh in our experience the effect of any degree of motion sickness. Besides, the excitement of the flight and to a certain degree the nervousness about performing well (and getting ill…) is controlled by the medication. A not too heavy dinner without alcohol, as much rest as possible during the night before the flight and a light breakfast are also strongly suggested to be in a good shape for the flight.

Both the effects of motion sickness medication as well as the likely minor effects of motion sickness most flyers experience emphasize the need for being well-prepared and practiced for the flight activities (i.e., being able to do them blind and sleeping). Being prepared involves two steps: first, ensure that the research you want to conduct is not affected by becoming ill. For MarsSedEx I and II, all critical activities were carried out during the first few parabolas when the risk of getting ill is low. However, if researchers have to perform for the entire duration of the flight, the possibility of being incapacitated by motion sickness should be accounted for. This is done best by being prepared to run the research activities being short of hands, for example, by keeping some parabolas free of regular activities, or taking an extra operator along. Taking a look around before parabolas start and considering where people rest and move is also useful to keep out of harms way and continue working even if somebody is ill next to you.

The second step to prepare for motion sickness, is to be ready for it, both becoming ill yourself as well as the person next to you. This involves having a space to retreat to when starting to feel ill. As soon as you retreat, indicate to your coresearchers, fellow flyers, and the flight crew that you do not feel well. Also, know where your sick bag is and keep it in a pocket of your flight suit that is easy to reach. One factor contributing to the risk of becoming ill is rapid body, especially head, movement when the inner ear and eyes are already sending conflicting signals to the brain (and stomach). Having to bend over to grab the sick bag in your lower leg pocket while bottoming out through a parabola at $2g$ or when zero g just sets in is certainly not helpful to avoid further deterioration of your condition, so keep it easy to reach. The time between starting to feel ill and vomiting is also known to be very short and vomiting can catch flyers by surprise. Retreating early to your safe space, indicating that you are not feeling well and grabbing your bag is therefore important to ensure that the situation is managed as well as possible. Motion sickness can hit anybody (even astronauts!), so it is nothing to be ashamed of and therefore no time to be heroic.

Apart from the motion sickness, several further points should be considered to ensure an adequate research performance during the flight. All flyers wear a flight suit similar to a military pilot's suit. The major benefit of the suit is all its pockets, which are very convenient to store bits and pieces of equipment. However, flight participants should keep in mind that packing too much into the suit limits the ability to move. Besides, one has to take the items out and put them back in to avoid floating away. Therefore, one should take only what is really required as a personal item or tool and fix everything else to the equipment you are using. Also, pack the suit pockets in a way that is practical for the situations you encounter during the flight. Less relevant items (flight socks, ID, mementos) can go into the lower leg pockets; pens, notebooks, and camera should be easy to reach in the belly pocket. Make sure you know where the individual items are because searching for them during parabolas is probably wasting valuable research time and may increase the risk of getting ill. With regards to clothes to wear under the suit, there are a few further considerations. The air-conditioning on board is limited and depending on the date and location of the flight, participants should also remember the likely change in temperatures when dressing for the flight. Our measurements during MarsSedEx I and II indicate lowest temperatures of approximately 15 °C (60 °F) in the cabin. We also recommend to wear shoes that can be changed easily because flyers take them off in exchange for flight socks after reaching cruising altitude and put them on again in preparation for landing, both while sitting in their seats. Apart from changing temperatures, the noise level in the cabin is also fairly high, so being able to communicate with signs or using headsets if a lot of talking is required is recommended. Finally, the lighting in the plane is rather low, so camera settings should be adjusted in order to account for the limited light.

Concluding these thoughts on the personal well-being and cabin environment we would like point out that the space given to a participant on a research flight is not designed to perform like Superman (or Kate Uptown in Sports Illustrated for that matter). While enjoying the experience is certainly what any participant should aim for, the unwritten etiquette demands that one must not impede the work of the fellow flyers by getting in their way (neither yourself nor your breakfast...image how Kate would look like if these bubbles were not water....).

BIBLIOGRAPHY

Roach, M., 2010. Packing for Mars. The Curious Science of Life in the Void. Oneworld, Oxford, p. 334.

Upton, K., 2014. Sports Illustrated, February 21, 2014.

Key Results of the MarsSedEx I Mission

ABSTRACT

The MarsSedEx I mission consisted of measurements of settling velocity and sorting of glass spheres on Earth and reduced lunar and Martian gravities. The results show that experiments on sedimentation on board-reduced gravity flights are possible. They also confirmed the critical review presented in Chapters 1, 4, and 5 of this book on the effect of reduced gravity on sediment sorting and settling velocity on Mars. In particular, the discrepancies between observations and the way settling sediment velocity is commonly modeled for planets other than Earth, indicate significant differences, which require consideration for the further exploration of sedimentary rocks, and thus search for traces of life, on Mars.

9.1 MarsSedEx I MISSION OBJECTIVES

The main objective of the MarsSedEx I flight was to assess the feasibility of short-duration experiments in the reduced gravity environment of parabolic flights. The two experiments presented in this book included settling tests using a one- and a three-chamber settling tube instrument. The OCST aimed at measuring the actual settling velocity of 1 mm spherical reference particles using a video recording of their settling velocity. The TCST enabled the measurement of the effects of reduced gravity on the sorting of sediment. For this test, a mixture of glass spheres ranging from nominally 212 to 1100 μm in diameter (Table 6.1) was used. Flight conditions on November 18, 2012 inhibited the completion of the full flight program so that a second flight was

Experiments in Reduced Gravity: Sediment Settling on Mars. DOI: 10.1016/B978-0-12-799965-4.00009-1

required on November 19, 2012. The first flight had been long enough to complete the MarsSedEx I measurements so that both tests could be repeated during the second flight. For the OCST, settling velocity could be recorded for lunar and Martian gravities, for the TCST a replicate test under lunar gravity appeared most feasible within the constraints on the operator during the flight. Figure 9.1 shows the g forces during the parabolas for OCST and TCST.

9.2 ONE-CHAMBER SETTLING TUBE RESULTS

The OCST enabled the observation of reference particles released from a holding chamber. The movement of the particles was recorded at 30 frames per second. After the flight, the movement of the particles in front of a ruler attached to the back of the settling tube was analyzed by counting the number of frames an individual particle required to move 1 cm downwards. The velocity was then calculated by

(i) Settling time in seconds for 1 cm = number of frames divided by 30

(ii) Settling velocity = 1 divided by the settling time calculated in (i)

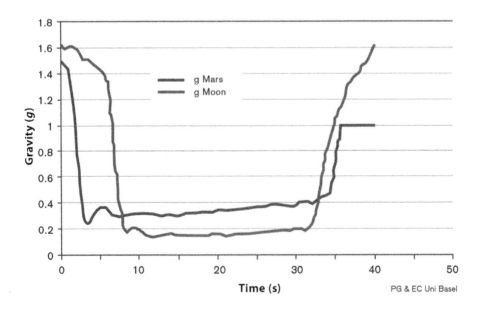

Fig. 9.1. g forces during the parabolas for one- and three-chamber settling tube tests.

To ensure that the terminal velocity had been reached, frames were counted for three adjacent 2-cm long measurement intervals along the entire settling path. The distortion due to the central perspective of the camera and the varying distance between particle and camera was assumed to be negligible due to the short length of the measurement interval compared to the distance of the camera and the particle (10 cm). Figure 9.2 shows a series of images with settling particles in

	Starting point	Same point 30 frames (0.25 sec) later	Distance (cm)
	15.25	17.4	2.15
	15.9	17.9	2
	15.3	17.1	1.8
	15.5	17.6	2.1
	15.1	17.5	2.4
		Mean (cm)	2.09
		Velocity (cm/sec)	8.36

Fig. 9.2. Reference spheres moving in reduced gravity and the associated settling velocities and gravities. The pairs of images show 10 marked particles moving along a section of 6 cm in the middle of the frames taken by the video camera. The section is split into three subsections of 2-cm length each to test whether particle-settling velocity was constant.

1-cm intervals. The table in Figure 9.2 lists the number of frames a particle required to cross a distance of 2 cm at the top, center, and bottom of the field of view of the recording camera. The velocities do not differ significantly, indicating that terminal velocity had been achieved and that the distortion of the image was negligible. The settling velocities illustrate that both duration of the test and the length of the settling tube were sufficient to achieve terminal settling velocity.

The achievement of this aim of the flight enabled a first and cautious analysis of the collected data with regards to the use of existing, semiempirical settling velocity models for Martian gravity. To do so, several comparisons of the combined settling velocity model and the observations for lunar, Martian, and terrestrial gravities were made (Tables 9.1 and 9.2). The observed settling velocities were significantly (t-test of means at with 5% probability of error) lower for Martian and lunar gravities than for Earth. The reduction was also greater than the ratio of gravities in the combined model, which suggests an approximate 50% decrease for Mars, instead of 60% as indicated by the difference in gravity terms in the combined model, and 29% for the Moon. This is a first indication that drag at reduced gravity cannot simply be implied using observations from Earth.

Table 9.1. Observed and Calculated Settling Velocities for MarsSedEx I and Settling Conditions

Test	Gravity (m/s²)	Kinematic viscosity (kg/m/s)	Observed settling velocity (cm/s)	Estimated settling velocity (cm/s)
Lunar	1.45	0.0000011198	5.73	4.47
Martian	3.24	0.0000012038	10.03	7.64
Earth	9.81	0.0000011528	20.07	15.93

Table 9.2. Calibration Data for MarsSedEx I

Test	Observed (cm/s)	Modeled (cm/s)	C_1	C_2
Moon	5.73	7.2	2.5	0.4
Moon		5.2	18	0.21
Moon		6.0	12	0.267
Mars	10.03	11.2	2.5	0.4
Mars		9.2	18	0.21
Mars		10.02	12	0.267

Comparing observed and calculated settling velocities (Table 9.1) indicates that the combined, noncalibrated model generally underestimates the settling velocities of the reference particles. We attribute this underprediction to the coating of the reference spheres, which presumably reduces the skin friction compared to spheres made of silicates. Therefore, calibrating the combined model based on the settling velocity observed on Earth is a sensible way of preparing the model for use on Mars. The calibration was done in three ways: by adjusting C_1, C_2 or both, to get a good prediction (<2% difference) for the settling velocity observed on Earth. To limit the number of C_1 and C_2 combinations, only one fit with a similar proportional change of both coefficients is shown here. These coefficients were then used to calculate the settling velocities measured during the two MarsSedEx I flights. Comparing the adjustments made to each coefficient and the quality of the prediction gives some insights into the limitations of the model.

Table 9.2 shows the results of the calibration. Generally, the predictions can be improved significantly by this calibration, almost achieving 99% when adjusting C_1 and C_2 at the same time by the same proportion. However, the adjustment raises several questions. First, both C_1 and C_2 were reduced to match observations on Earth. This is not plausible because C_2 mostly covers the effect of form on drag and is set at 0.4 for spheres, increasing for less regularly shaped particles. The increase proposed in the literature for natural grains makes physical sense because the irregular shape increases form drag. We speculate that the reduction required for our reference particles is a consequence of several differences during our experiments compared to the tests conducted for the development of the combined model. First, the reference spheres do have a lower density and the temperatures for our tests were several degrees lower than most used to derive C_1 and C_2 in the literature. Both factors may be beyond the boundary conditions for the model. Tests conducted with reference spheres (Table 9.3) showed the sensitivity of settling velocity to temperature. Generally, the model reflected the effect of declining kinematic viscosity, which should lead to slower settling. However, the error for the noncalibrated model ranged from 17% to 27% and increased with rising increasing water temperature. For the calibrated model, i.e., adjusting both C_1 and C_2, the effect of lower temperatures was overpredicted, while the increase of settling velocity with

Table 9.3. Water Temperature and Settling Velocity of MarsSedEx I OCST Spheres on Earth

Temperature (°C)	Settling velocity (cm/s)	Difference observed – predicted for Earth
13.8	19	−17%
16.7	20.07	−
23.2	22.95	−27%

water temperature was not fully reflected by the model. This implies that the model works well at water temperatures close to those at which C_1 and C_2 values were set. The settling velocity on Earth used for calibration was selected accordingly, matching the one observed during the flight most closely.

Comparing best fits for changing only one of the two coefficients still offers some insights into the applicability of the model under reduced gravity. Leaving C_2 at 0.4 generates a best fit at a C_1 of 2.5. Reducing the effect of viscous drag, however, leads to poor prediction results for lunar and Martian gravities, overestimating the settling velocity by 12% (Mars) and 27% (Moon). This error is plausible because viscous drag should have a stronger influence at reduced gravity and lower settling velocities. The most sensitive calibration therefore should be a reduction of C_2 compared to C_1. However, adjusting C_2 or both parameters to achieve the best fit for Earth carries the risk of getting good results for the wrong reasons. Reducing C_2 below a value of 0.4 reported in the literature for spheres is not physically plausible. We still considered it acceptable in this first test because a key outcome, which is physically plausible, is the limited ability of the semiempirical combined model to properly reflect the ratio of viscous to turbulent drag at lower terminal settling velocities. An appropriate correction would involve the development of a correction factor for gravity in the turbulent drag term of the equation. Reducing C_2 to test whether this generates a better fit has mathematically the same effect. Therefore, it was considered to be an acceptable approach for the one set of particles used in this test.

Reducing C_2 to get a good fit for Earth produces an average underprediction of 8% for both Mars and Moon. This is plausible because turbulent drag is smaller at reduced gravity. The result of the calibration of C_1 and C_2 is the best of all, with less than 1% error for Mars and 5%

for lunar gravities. In general, these small errors can be considered a good result, indicating that the combined model would be suitable in principle to further explore settling velocities and sediment sorting on Mars when based on a calibration.

The potential for extrapolation of the encouraging result achieved by the calibration of C_1 and C_2 can be checked by comparing the quality of the C_1 and C_2 coefficients in the equation derived for C_d based on the combined model with

$$C_d = \left(\frac{2C_1 v}{\sqrt{3RgD^3}} + \sqrt{C_2} \right)^2 \qquad \text{Equation 9.1}$$

Unfortunately, the drag coefficients calculated with each combination of C_1 and C_2 in Table 9.2 are consistently lower than those based on observations (Table 9.4). The difference is on the order of 40–60%. Consequently, fitting C_1 and C_2 for the drag coefficient equation above derived from the combined model to match the observed C_d, generates C_1 and C_2 values of 24 and 0.67, respectively, much greater than those for a good prediction of settling velocities. This large discrepancy between the two, supposedly related equations, can therefore be seen as a confirmation of the risk of using the combined model outside the boundary conditions it was developed for without due calibration.

Table 9.4. Observed Reynolds Number and Observed (Based on the Equation for $20 \leq Re \leq 260$ in Appendix II) and Estimated Drag Coefficients (Using Equation 9.1) for MarsSedEx I OCST Particles

Test	Particle Reynolds number	Observed drag coefficient	Drag coefficient based on calibrated combined model	
Moon	52	1.53	$C_1 = 2.5, C_2 = 0.4$	0.5
			$C_1 = 18, C_2 = 0.21$	0.95
			$C_1 = 12, C_2 = 0.267$	0.74
Mars	81	1.21	$C_1 = 2.5, C_2 = 0.4$	0.47
			$C_1 = 18, C_2 = 0.21$	0.67
			$C_1 = 12, C_2 = 0.267$	0.58
Earth	178	0.82	$C_1 = 2.5, C_2 = 0.4$	0.44
			$C_1 = 18, C_2 = 0.21$	0.44
			$C_1 = 12, C_2 = 0.267$	0.43

9.3 THREE-CHAMBER SETTLING TUBE RESULTS

The TCST was conducted twice during lunar gravity on November 18 and 19, 2012. Due to the limited time for cleaning and refilling the instrument after the first flight some contamination of particles may have occurred. Therefore, only results of the November 18 flight are reported here. The settling time of the particles was 15 s. After the flight, the top, middle, and bottom chambers of the TCST were emptied and the particle size distribution for the sample from each chamber was measured using a Malvern Mastersizer in Basel. However, even the interpretation of these results has to be done with some caution because many small glass spheres stayed attached to parts of the tube during emptying. Weighing the retrieved sediment back in Basel showed that about 20% of the sample had been lost. We attribute this to the electrostatic forces between the particles and the settling tube. The limited facilities to clean the settling tube chambers, first during the unexpected reloading after the November 18 flight and then in the hotel after the second flight, was the reason for the partial loss of the sample. Further errors were introduced because the samples were decanted, transferred into a travel safe container, and then air-dried in a hotel room to ensure they could be carried back to the Basel labs in carry-on luggage. In Basel, further transfers from travel-containers to analytical beakers and the Mastersizer contributed to the losses. Despite this problem, which presumably affected smaller particles to a greater extent than the larger ones, we feel that, at least, the first test generated data that are distinct enough to present.

Figure 9.3 A to C shows the cumulative distribution of particles in the TCST for both terrestrial and lunar gravities for the three TCST chambers. The shift of particle size distribution between the original mixture and the three chambers on Earth illustrates the effect of the different settling velocities of the particles. The large ones settle faster and more of them reach the bottom chamber. Fine particles, on the other hand, are depleted. This effect occurs both for terrestrial as well as lunar gravities. However, it is less pronounced for the lower gravity, i.e., the difference between the original sediment mixture and the size distribution in all chambers is less pronounced for the reduced gravity. The less-pronounced sorting is attributed to the lower lunar settling velocities. The lower settling velocities coincide with smaller settling velocity

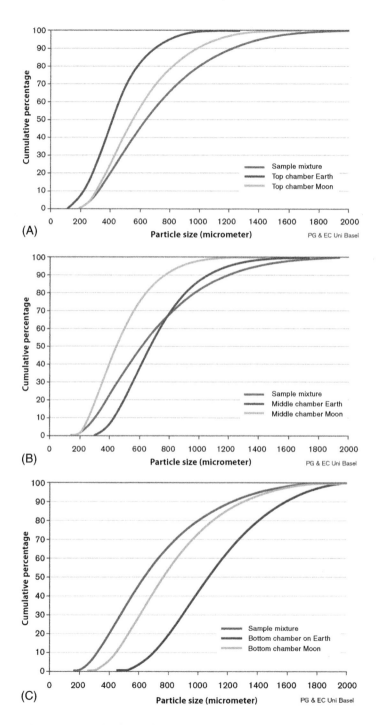

Fig. 9.3. Cumulative particle size distribution of the TCST particles after settling on Earth and Mars from top (A), middle (C), and bottom (C) chamber.

differences. This, in turn, leads to smaller distances between particles along a given travel path because the faster settling particles advance less compared to Earth. The data confirm the relevance of theoretical considerations based on the effect of gravity on sorting. Figure 9.4 shows the distances traveled by particles ranging from silt to sand size in 15 s using the noncalibrated combined model. Settling velocities decline with size, but to a larger degree for Earth than lunar gravity, illustrating that fine sands would be less sorted when deposited from a given, shallow water layer.

9.4 CONCLUSIONS FROM MarsSedEx I FLIGHTS

The comparison of noncalibrated settling velocity model outputs and data observed for MarsSedEx I reveals an underestimation at reduced gravity. Calibration for Earth led, to a lesser degree, to an overprediction.

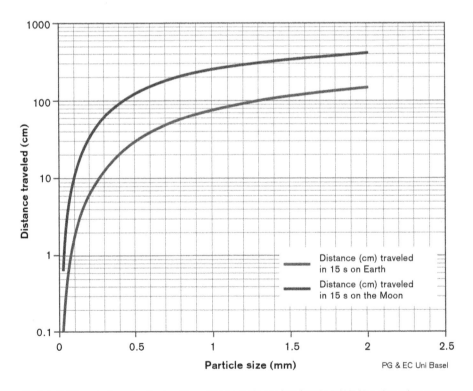

Fig. 9.4. Settling velocity and sorting in a 15-cm deep layer of water based on the TCST. Note the nonlinear increase of settling distance with particle size indicating that for sediment in this size range the sorting would be poorer on Mars than on Earth.

While these errors are physically plausible, they nonetheless highlight the risk of applying a semiempirical model outside a calibrated realm. Combining results of OCST and TCST indicates that applying the degree of sorting observed on Earth to Mars may lead to a biased assessment of the flow conditions that generated the sediment. On Mars, lower settling velocities resulted in less-pronounced sorting from a still body of water of a given depth. In moving water, the effect is maintained because lower gravity also causes lower flow velocities, so that the distance a particle travels while settling is also shorter, contributing to less sorting. At given runoff rates, flow on Mars is deeper than on Earth, which may balance the effect of gravity on sorting, but the effect requires a quantitative assessment similar to those conducted for terminal velocities. Overall, the very preliminary results of MarsSedEx I and II indicate that caution is required when using Earth analogs of sediments to assess flow hydraulics and thus, the sedimentary environment on Mars. Sediments containing most traces of life on Earth are usually fairly well sorted and fine grained. The lower degree of sorting expected on Mars, as illustrated by our tests, highlights the need to consider sediment layers with less sorting and possibly, a large fraction of sand as still potentially bearing traces of life. MarsSedEx I also showed that conducting settling experiments under the technically and environmentally restraining conditions of a reduced gravity flight is possible.

BIBLIOGRAPHY

Dietrich, W.E., 1982. Settling velocity of natural particles. Water Resources Res. 18, 6.

Ferguson, R.I., Church, M., 2004. A simple universal equation for grain settling velocity. J. Sedimentary Res. 74, 5.

Grotzinger, J.P., Hayes, A.G., Lamb, M.P., McLennan, S.M., 2013. Sedimentary Processes on Earth, Mars, Titan, and Venus. In: Mackwell, S.J. et al. (Ed.), Comparative Climatology of Terrestrial Planets. University of Arizona, Tucson, pp. 439–472.

Julien, P.Y., 2010. Erosion and Sedimentation, Second Edition. Cambridge University Press, Cambridge, p. 371.

Knighton, D., 2014. Fluvial Forms and Processes: A New Perspective, Second Edition. Routledge, New York, p. 383.

Komar, P.D., 1979. Modes of sediment transport in channelized water flows with ramifications to the erosion of the Martian outflow channels. Icarus 42, 13.

MarsSedEx II

ABSTRACT

The successful completion of the MarsSedEx I mission showed the need for more information on the effects of gravity on particle settling and the ability of simple models developed on Earth for application on Mars. MarsSedEx II consequently focused measuring the settling velocities of a range of particles with an increasing resemblance to sediment that may be found on Mars. For this mission, the flight hardware was modified based on the experience of MarsSedEx I to maximize the data output. The following chapter reports the rationale for the sediment selection, presents details on the hardware modification, and reflects on the practical aspects of conducting the experiments during the reduced gravity flight in November 2013.

10.1 AIMS AND OBJECTIVES OF MarsSedEx II

MarsSedEx I had shown that the measurement of terminal settling velocity and sediment sorting is possible under reduced gravity conditions. In addition, the flight also delivered a scientific result by illustrating that the expected reduction of drag for Moon and Mars was not reflected by the drag parameter values, both noncalibrated and adjusted to observations on Earth. However, all measurements had been conducted with coated glass spheres and not real sediment. The TCST test had also been problematic, because some sediment was lost, presumably due to the limited facilities or cleaning the tube in the plane

Experiments in Reduced Gravity: Sediment Settling on Mars. DOI: 10.1016/B978-0-12-799965-4.00010-8

and a hotel room as well as losses during the transfer from Florida to Switzerland. The aim of the MarsSedEx II mission, i.e., flight and related experiments on the ground, was to test the use of the combined settling velocity model for estimating terminal settling velocities of real sediment particles under Martian gravity. The focus of the MarsSedEx II flight was therefore on measuring settling velocities of particles with an increasing resemblance to real sediment, including differences in its critical properties such as size and shape. A new experiment on measuring the effects of gravity on the sorting of sediment will be designed for MarsSedEx III.

The objectives of MarsSedEx II were to:

1. measure the terminal settling velocity of a range of natural sediment particles for terrestrial and Martian gravities;
2. calibrate the semiempirical combined settling velocity model for Earth; and
3. test the applicability of the terrestrial calibration on Mars.

To achieve these aims, instruments were modified based on the experience and performance of MarsSedEx I.

10.2 DEVELOPMENT OF MarsSedEx II INSTRUMENTS

MarsSedEx II was to use the successful one-chamber tube method developed for MarsSedEx I for a wider range of sediment particles. Focusing on one type of test with a similar instrument kept changes to the instrument-carrying structure compared to MarsSedEx I at a minimum. Apart from reducing the design, testing and construction effort, this also facilitated the preparation of the Research Proposal to Zero-G as well as the Flight Readiness Review. While the same basic frame design was used for the instrument-carrying structure, three major modifications were carried out (Figure 10.1). The first was the redesign of the two endplates. Their shape was changed by having a semicircular cutout on each side. This modification aimed at facilitating the access to the settling tubes, in particular, the handles of the ball valves. The second change to the endplates was the cutting of slits into each corner to hold the belts fixing the apparatus to the floor. Finally, to stabilize the structure and to enable the mounting of cameras, two 10-mm diameter, upright rods

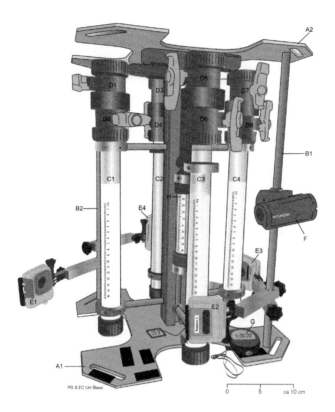

Fig. 10.1. Key components of the MarsSedEx II instrument. A1 and A2: Bottom and top endplates; B1 and B2: upright supports; C1–C4: settling tubes 1–4; D1–D8: ball valves and sediment-containing chambers; E1–E4: cameras 1–4 for recording of sediment settling; F: monitoring camera required for synchronization; G: stopwatch for recording experiment time; H: short tube with thermometer.

were mounted in two opposite corners (…when the mounting belts were wrenched tight the day before the flight, I wished there was one in each corner, and nightmares of a broken endplate dominated the little sleep the author had the night before the flight…) (Figure 10.2).

The settling tubes were also modified from the successful MarsSedEx I design. The settling velocities observed during the lunar and Martian gravities of MarsSedEx I had shown that a settling distance of 35-cm length was sufficient to achieve terminal velocity and offer a long enough section to record it with an action cam. To take advantage of this shorter tube requirement, two supply chambers were mounted on the top of the tube. The chambers themselves were the same ball valves as those used for MarsSedEx I, connected by a custom-made

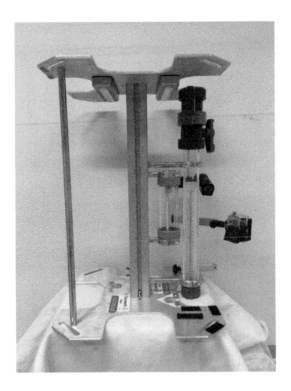

Fig. 10.2. MarsSedEx II instrument frame with settling tubes in a laboratory configuration with LED lighting packs attached to the top plate and short tube for water temperature recording. Credit: B. Kuhn.

plastic threaded plastic tube. Releasing particles after each other by first opening the lower rather than the top chamber doubled the amount of data that could be collected from a single tube. Overall, four tubes and eight chambers were available for MarsSedEx II. A fifth, short tube was mounted between the central upright beam and one of the settling tubes. This short tube contained only water, a thermometer, and a temperature logger. The addition of this short chamber was a consequence of the temperature drop observed in water between FRR, mounting and flight.

Apart from the instrument-carrying frame and the settling tubes, the equipment used in the periphery of the settling tubes was also modified based on the MarsSedEx I experience and the objectives of MarsSedEx II. To measure gravity, the same MSR-145 logger and an iphone 4 App called gmeter were used. The cameras, now five, were changed to GoPro Silver action cams and one Hyundai Speedcam Full HD. The main reasons for the choice of cameras were the quality of the images they delivered, the increased number of frames per second as well as their reliability. Furthermore, five cameras with monitors and cables would have cluttered the instrument-carrying frame. None of the cameras has a display, so the accurate positioning had to be checked with a marked ruler while starting up the instruments during preparabola cruise flight. The settling tubes were monitored by a GoPro Silver each, recording at 120 frames per second (fps), compared to 30 fps for MarsSedEx I. The Hyundai Speedcam Full HD was used to record the gmeter app reading displayed on the iPhone and a stopwatch recording the flight time. The watch in this camera was synchronized with the MSR-145 g logger. This, admittedly somewhat crude, synchronization of recording devices enabled a more accurate identification of the actual gravity while particles were moving than during MarsSedEx I. The main reason for introducing the advanced synchronization of recording devices was the variability of the gravity during the flight (Figure 10.3). The more accurately the gravity between release and recording of particles is known, the better the calibration of the model.

A further addition to the instrument suite were two rows of battery-powered LEDs to improve the lighting conditions. The cases containing

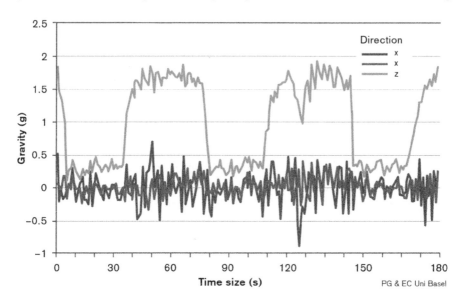

Fig. 10.3. *Gravity during the MarsSedEx II Martian parabolas. The relevant values are the three lower intervals of Z. All settling observations were carried out during the first 5–10 s of the parabolas. Average g values for these intervals were used for further calculations.*

LEDs and batteries were mounted onto the bottom of the top endplate. They improved the quality of the video clips significantly. However, lighting was still not ideal and remains a concern for future missions. The final peripheral devices that were used consisted of a Microlite LITE5008P-A USB temperature logger and the aquarium thermometer bought for MarsSedEx I. Both were placed in the small tube for water temperature measurements. The old aquarium thermometer saved the experiment because the cap covering the USB connection of the temperature logger turned out to be not water tight, flooding the logger so that it recorded no valid numbers. Always carry a backup!

10.3 SEDIMENT PARTICLE SELECTION

After MarsSedEx I, it would appear most obvious to test sediment looking as "Martian" as possible right away. However, this approach lacks an "understood" reference because sediment particles have a natural shape. Our understanding of the interaction between such complex shapes and settling velocity is limited, therefore, a stepwise approach to accommodate the increased complexity of particle shape was taken. To achieve the objectives of MarsSedEx II with a number of tests limited to eight chambers, the selection of the particles aimed at a gradual transition from the known 1-mm reference spheres to real sediment with properties as similar as possible to sand encountered on Mars. Therefore, a wider range of sizes and shapes of particles was tested to fully explore the transition between viscous and turbulent drag. Furthermore, practical criteria played a similarly important role as the resemblance to possible Martian sediment. All particles had to be large enough to be visible on the recordings made by the cameras. Moreover, they should differ in settling velocity and enable a test of the quality of the combined model for assessing settling velocities on Mars. The properties of the sediment particles used for MarsSedEx II are listed in Table 10.1. Close up images of the particles are presented in Figure 10.4.

Four specific comparisons to test the ability of the combined model to describe the effect of sediment properties on settling velocities were conducted. First, basalt spheres of 1-mm diameter were compared to the MarsSedEx I spheres, referred to as reference spheres in this section. This comparison aimed at testing the effect of density on settling

Table 10.1. Properties of the Particles Tested During MarsSedEx II				
Sample	Material	Size/settling velocity	Density (g/cm³)	Origin
Pink spheres	Glass microbeads, Fuchsia C-MGB-FC1000	1000–1180 μm, D_{50} 1020 μm	2.33	Microspheres–nanospheres Cat. No: 155833-50
Small basalt spheres	Basalt	500–600 μm, D_{50} 550 μm	2.75	Whitehouse Inc.
Medium basalt spheres	Basalt	710–850 μm, D_{50} 760 μm	2.75	Whitehouse Inc.
Large basalt spheres	Basalt	800–1000 μm, D_{50} 1000 μm	2.63	Whitehouse Inc.
Slow basalt sand	Nephelinite	9.5–12 cm/s, D_{50} 1003 μm	2.53	Maarmuseum Manderscheid
Medium basalt sand	Nephelinite	12–16 cm/s, D_{50} 1003 μm	2.67	Maarmuseum Manderscheid
Fast basalt sand	Nephelinite	16–25 cm/s, D_{50} 1015 μm	2.77	Maarmuseum Manderscheid
Amazon sand	Amazon Green Sand, Quartz	1–2 mm, D_{90} 1400 μm	2.5	Qualipet, Switzerland
Hematite sand	Hematite placer	16–25 cm/s, D_{50} 800 μm	4.3	Corsica, collected by H.-R. Rueegg

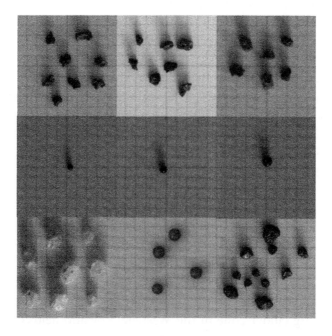

Fig. 10.4. Sediment particles used for MarsSedEx II. Most notable differences, apart from size, are the degree of rounding, and the heterogeneity of mineralogy. Top row: basalt sand; left to right: slow, medium, and fast settling; middle row: basalt spheres 500, 700, and 1000 μm; bottom row, left to right: Amazon sand, reference spheres, hematite. Credit: B. Kuhn.

velocity and the general replicability of reduced gravity tests. The second comparison used *basalt spheres* of 500–600, 710–850, and 800–1000 μm manufactured by Whitehouse Scientific Inc. to test the effect of size, and thus the ratio between viscous and turbulent drag. These spheres have the density of natural silicate sediment particles and their change in size enables a test of the quality of the calibration of the C_1 coefficient for viscous drag in the combined settling velocity model for a range of particle sizes.

The third test conducted during MarsSedEx II focused on particle shape. Four particles of roughly the same diameter were tested. Apart from the basalt sphere, well-rounded quartz sand, and increasingly angular *hematite* and *basalt sands*, indicated by the increasing specific surface area in Table 10.1, were used. The well-rounded quartz sand, referred to, after its product name, as *Amazon sand* enables the comparison of the MarsSedEx II data to the C_2 factor values describing shape reported in the literature. To narrow the range of the particle size to an equivalent quartz size (EQS) (see subsequent sections for details) of 1 mm, the naturally green-colored sand sold for furnishing terraria was fractionated to settling velocities between 16 and 25 cm/s by settling tube. This fairly wide range was determined by the sample size that could be fractionated with the large Basel settling tube. Still, this particle turned out to be difficult to identify on the videos taken during the MarsSedEx II flight. Comparing the number of particles that could be identified suggested that only the largest Amazon sand was visible. Accordingly, the D_{90} rather than the D_{50} was used as an appropriate measure of diameter. The third particle selected for the roundness test was an angular, dense, and fast-settling hematite. Density and shape of this particle were quite outside the realm of the other sediments, but it was selected on purpose to enable a sensitivity test of the calibrated combined settling velocity model for such an extreme shape and density. The hematite used for MarsSedEx II was also fractionated by settling tube to velocities between 16 and 25 cm/s. The final particle selected for this test was basalt sand. This enabled moving further toward a Martian sediment. The sand forms a colluvium that originates from a basaltic ash generated by the Mosenberg cinder cone in the Westeifel volcano field, Germany. It has the density of a silicate, but is much more angular than the Amazon quartz. Particles also had a settling velocity of 16–25 cm/s.

The fourth test conducted during MarsSedEx II focused on the differences between settling velocity of a wider range of basalt sand particles. In addition to the *fast-settling basalt sand* used for the roundness test, two slower settling fractions of the same basalt ash were measured. The ash particles were dry-sieved to a size ranging from 0.5 to 2 mm. Further fractionation was not done by size, but according to settling velocities of basalt spheres of 0.5, 0.75, and 1 mm using the 1.8-m Basel settling tube, corresponding to the concept of EQS. Based on the settling velocities of the reference spheres observed during MarsSedEx I and a conservative estimate of the duration of reduced gravity, a range of settling velocities for Mars was estimated to fractionate the sample material into subsamples of a suitable, narrow range of values that would generate clearly distinguishable particle clouds. This fractionation, rather than size, also enabled a comparison of form on settling velocity when comparing the basalt ash to the basalt spheres as well as the ability of the combined model to estimate the actual settling velocity.

In total, nine samples were selected for MarsSedEx II. This is one more than the eight chambers available on the four MarsSedEx II settling tubes can hold. The preparatory tests had revealed that particles could be distinguished based on their color on the video clips showing the settling sediment. Therefore, some particles were mixed with the pink reference spheres into one supply chamber (Table 10.2). This also generated a comparison of the effect of potential differences between settling in the four tubes during the different parabolas flown for MarsSedEx II.

Table 10.2. Distribution of Sediment Particles in the MarsSedEx II Settling Tubes								
	Tube 1		Tube 2		Tube 3		Tube 4	
Sample	Bottom	Top	Bottom	Top	Bottom	Top	Bottom	Top
Pink reference	X	X					X	X
Large basalt spheres	X							
Medium basalt spheres					X			
Small basalt spheres						X		
Green sand		X						
Hematite sand								X
Fast basalt ash							X	
Medium basalt ash		X	X					
Slow basalt ash				X				

10.4 SET-UP AND FLIGHT PLAN OF THE MarsSedEx II

The general setup of the MarsSedEx II settling tubes did not differ much from MarsSedEx I (Figure 10.5). The distribution (Table 10.2) of the tested samples into the supply chambers aimed at ensuring the best visibility of the settling sediment. The pairs of samples in the top and bottom chamber of each tube were expected to differ in settling velocity. Samples in the bottom chamber were always assumed to move faster than those in the top chamber. This way, the sample from the bottom chamber that was opened first moved along the tube quickly, getting out of the way of the following sample from the top chamber. The top chamber was supposed to be opened when the spheres from the

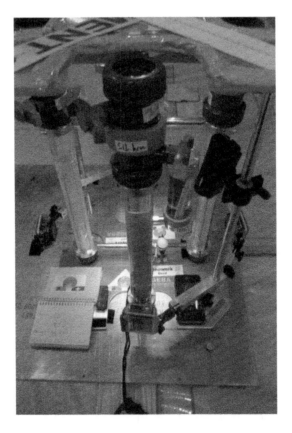

Fig. 10.5. The MarsSedEx II apparatus onboard G-Force One, ready for the flight as seen from the position of the operator. Note the additional instruments not used in the lab: iPhone with gmeter app running and MSR-145 g-logger, next to the flight plan in the near left corner. The astronaut figure serves as a lucky charm, an essential part of all experimental research. Credit: N.J. Kuhn.

bottom chamber passed the section of the ruler filmed for measuring terminal velocities.

Appendix IV contains a transcript of the actual MarsSedEx II flight plan. Due to the smaller number of different experiments and by considering the experience made during MarsSedEx I, it reads fairly straight forward. The only issue that arose for the operator was the difficult access to the ball valve handles of the tube mounted opposite the seating position of the operator. This potential problem could be addressed because during MarsSedEx II four parabolas of Martian gravity were flown. The tube opposite the operator was opened during the first parabola using one hand placed on the handle of the bottom and top chamber, respectively, before reduced gravity set in. The handles of the tubes on the left side and in front of the mounting frame (as seen by the operator) were easy to reach and could be opened with one hand each during one parabola. Accordingly, they were opened during the second parabola. Measurements with the fourth tube were done during the third parabola. This order of operating the tubes left the fourth Martian gravity parabola as a spare in case of delays.

10.5 PROBLEMS DURING THE MarsSedEx II FLIGHT

By and large, the flight plan for MarsSedEx II worked well. The only difficulties that arose were mostly operator errors. The first error occurred because the stopwatch showing the nominal flight time was started too late, after the LEDs had been switched on. The reason for a set procedure of switching on the recording devices is the lack of an internal clock of the GoPro Silver cameras. The Hyundai camera has such a clock, showing the time on the recorded clips. The MSR 145 g logger and the Hyundai camera were set at the same time. Switching on the LEDs after all cameras and the stopwatch were running, was supposed to be a signal recorded by the GoPros that could be used to synchronize all recording devices. The error in the order of switching on stopwatch and LED occurred because the flight plan had two activities noted under the same item number. Focused on setting up the cameras, i.e., the main recording device, properly, the operator skipped the latter instruction and the stopwatch was started later on. Fortunately, a series of other movements could be used for synchronizing all devices, and thus the errors caused did not impede on the quality of the data collected during the flight.

The second error was also due to a combination of operator performance and the instructions in the flight plan. The time difference between opening the bottom and top chambers during the first parabola was very short, generating only a small gap between the particle clouds. The problem is attributed to the complexity of the situation. The flight plan did not contain an instruction on how long to wait between opening the chambers, such as "count one-thousand, two-thousand, three-thousand." Furthermore, the settling sediment was barely visible to the operator sitting in front of the instrument frame, with both arms restricted to the inside, having to move handles in the right order, while trying not to incur weightlessness, and checking gravity on the iPhone app during the first parabola of the flight. This all caused a limited sense for good timing and the nerves of the first parabola got the better of the operator and the gap between sediment clouds was very, but fortunately not too short. Both errors emphasized that while the flight plan had clearly improved between MarsSedEx I and II, it is also obvious that the situation demands a detailed flight plan and enough practice before the flight, as well as a reassessment of the plan after FRR and mounting.

BIBLIOGRAPHY

Ferguson, R.I., Church, M., 2004. A simple universal equation for grain settling velocity. J. Sedimentary Res. 74, 5.

Grotzinger, J.P., Hayes, A.G., Lamb, M.P., McLennan, S.M., 2013. Sedimentary Processes on Earth, Mars, Titan, and Venus. In: Mackwell, S.J. et al. (Ed.), Comparative Climatology of Terrestrial Planets. University of Arizona, Tucson, pp. 439–472.

INTERNET RESOURCES

http://www.sandatlas.org/

CHAPTER 11

MarsSedEx II Results

ABSTRACT

In this chapter the results of measuring settling velocities of sediment particles of natural shapes and lithologies similar to Mars on Earth and reduced gravities are reported. They confirm the need for recalibrating models developed on Earth and, more importantly, identify the basic problems associated with using an empirical model based on observations on Earth to a planet with a vastly different gravity. The results also confirm that the search for traces of life in sedimentary rocks should be based on a more precise numerical modeling than noncalibrated empirical models developed on Earth provide.

11.1 DETERMINING MODEL OUTPUT QUALITY FOR REAL SEDIMENT

MarsSedEx II tested whether simulating sedimentation on Earth with real sediment can be seen as an analog for Mars. Accordingly, the overarching aim of the MarsSedEx II flight and corresponding ground experiments was to determine whether a semiempirical model developed

Experiments in Reduced Gravity: Sediment Settling on Mars. DOI: 10.1016/B978-0-12-799965-4.00011-X

to describe settling velocity on Earth could be used to accurately predict settling velocities observed during reduced gravity. The ground experiments were conducted in the Planetary Surface Process Labs at the University of Basel, Switzerland, and the reduced gravity observations during a reduced gravity research flight operated by Zero-G on November 17, 2013 from Titusville, Florida.

A summary of the MarsSedEx II gravities, kinematic viscosities, and observed settling velocities is given in Tables 11.1 and 11.2. The quality of the output for varying model parameters was tested by comparing observed settling velocities to the results of a noncalibrated model and a calibration based on settling velocities observed on Earth. The

Table 11.1. Summary Results for Reference Spheres Observed on Earth and MarsSedEx I and II

Sample	Observed settling velocity (cm/s)	Gravity (cm/s²)	Kinematic viscosity (kg/m/s)	Particle Reynolds number
MarsSedEx II Tube 1 bottom	10.16	3.47	0.0000010040	103
MarsSedEx II Tube 1 top	10.14	3.33	0.0000010040	103
MarsSedEx II Tube 4 bottom	11.16	3.47	0.0000010040	113
MarsSedEx II Tube 4 top	11.62	3.28	0.0000010040	118
MarsSedEx 1 Moon	5.73	1.45	0.0000011198	52
MarsSedEx 1 Mars	10.03	3.24	0.0000012038	85
Earth	20.07	9.81	0.0000011528	178

Table 11.2. Observed Settling Velocities and Conditions for MarsSedEx II

Sample	Observed settling velocity Earth (cm/s)	Observed settling velocity Mars (cm/s)	Reduced gravity (m/s²)
Basalt sphere 500 μm	14.55	5.58	2.77
Basalt sphere 700 μm	18.07	7.26	3.29
Basalt sphere 1000 μm	20.23	8.46	3.47
Amazon sand	20.81	11.7	3.33
Hematite sand	22.4	14.94	3.78
Slow settling basalt sand	14.9	6.28	2.86
Medium settling basalt sand	16.18	5.9	3.17
Fast settling basalt sand	17.96	7.7	3.47
Kinematic viscosity Earth	0.0000009548 kg/m/s		
Kinematic viscosity Mars	0.000001004 kg/m/s		

analysis of the MarsSedEx II data followed a similar procedure for each variable of interest. First, the terminal settling velocities of the reference spheres were compared to those of Earth and MarsSedEx I. This enabled an assessment of the overall comparability of the flight conditions and further tests of the scientific results of MarsSedEx I. Further comparisons included the differences between Earth and Mars for particles of different density (reference particles versus 1-mm diameter basalt spheres), size (all basalt spheres), roundness (basalt spheres, Amazon, basalt, and hematite sand, all of approximately 1 mm in diameter), as well as settling velocity on Earth (fast-, medium-, and slow-settling basalt sands).

11.2 MEASUREMENT OF SETTLING VELOCITIES FOR MarsSedEx II

The general procedure used to determine settling velocity was not changed from MarsSedEx I. However, the video clips taken with the GoPro Hero 3 Silver cameras at 120 frames per second during the MarsSedEx II flight turned out to be of lower lighting quality than expected from tests on Earth and the MarsSedEx I flight. Therefore, a short description of the procedures used to generate suitable imagery for settling velocity measurements is added here.

The raw mp4 files taken during the flight were trimmed to the sections with settling particles using QuickTime to maintain the number of frames per second. The trimmed file was imported into iMovie to enhance shade, lighting, and color of the entire clip for easy particle recognition. iMovie was also utilized to select a shorter sequence for frame-by-frame analysis of the moving particles used for settling velocity determination. These video snapshots were selected in a way that showed at least 10 particles that could be clearly identified at a time moving through the measurement section. Subsequently, VLC was used to take a video snapshot of single frames showing individual particles at the start and end points of their path through the measurement section. These images were then enhanced with Pixelmator, which also served to mark the particles in the images with the MarsSedEx I and II results presented here (see Figure 9.2). For MarsSedEx II ground measurements, the number of frames a particle required to cross a fixed section of 2 cm length was counted to determine settling velocities. Due to the limited lighting of the flight clips, the length of the section was adjusted for the reduced gravity data. Counting still took place in the center of the recorded image, but instead of using fixed marks 2 cm apart, sections were selected when a particle was clearly visible for at least 20–30 consecutive frames. At settling velocities of approximately 10 cm/s, this number of frames corresponds to a similar distance. Generally, particles in the first half and the center of the moving cloud were counted, again to ensure good visibility. These particles are not moving faster than the rest of the cloud because the distribution is largely determined by the duration of the time required to open the ball valve, which stretches the release time to about a second, compared to 2 s for the entire cloud to move through the counted section. Despite these efforts to make out particles that moved clearly between two marks, the Amazon sand was difficult to identify on the flight videos. Only the larger particles could be clearly identified, which led to an adjustment of the particle size used for the calculations.

11.3 REPLICABILITY OF REDUCED GRAVITY EXPERIMENTS

Before testing real sediment, the replicability of the experiments conducted during MarsSedEx I in reduced gravity was checked by comparing reference spheres settling velocities observed during both research flights. Table 11.1 lists the settling conditions and observed velocities for the reference spheres during MarsSedEx II and for comparison the same information from Earth and MarsSedEx I is included. There was no significant difference between MarsSedEx I and II for Martian gravity (*t*-test of means at 5% probability of error). The settling velocities observed during MarsSedEx II are slightly greater than those of MarsSedEx I, which is attributed to the greater acceleration by gravity and slightly warmer water (20 °C) and thus, reduced viscous friction during the second flight. Comparing observations with predictions using both the noncalibrated and Earth-calibrated combined model supports the findings of MarsSedEx I (Figure 11.1). The noncalibrated model underpredicts the observed settling velocities, on average by 22%, while a good fit can be achieved with the model calibrated for Earth in Chapter 9. This good agreement is encouraging because it shows that settling velocity tests during reduced gravity flights are actually replicable and that the combined model can be used if calibrated for particles resembling those on Mars.

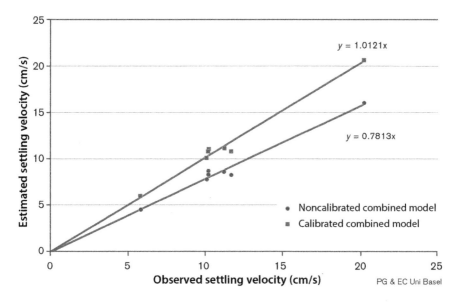

Fig. 11.1. Observed and estimated settling velocities for reference spheres observed at terrestrial, lunar, and Martian gravities. The r² for the calibrated model is 0.97.

11.4 DENSITY OF SPHERICAL PARTICLES

Comparing the reference spheres and the basalt spheres of 1-mm diameter gives an indication of how well the combined model covers differences in density and possibly skin friction between two particles of the same D_{50}. The MarsSedEx II results are listed in Table 11.2. Overall, the basalt spheres are slower than the reference spheres, which is somewhat surprising given their greater density and supposedly similar size. However, as discussed in Chapter 9, the coating of the reference spheres may reduce their skin friction compared to the basalt spheres, thus increasing their settling velocity, especially in flow conditions at the transition between Stokes and Newtonian. While this limits the assessment on the ability of the model to capture density differences, it enables testing the quality of adjustments of the coefficient C_1 determining viscous drag.

Using the noncalibrated combined model to calculate the velocities reveals an underprediction for the reference particles on both Earth and Mars of a similar order than for MarsSedEx I (Figure 11.2). For the basalt spheres, Mars is overpredicted and Earth underpredicted,

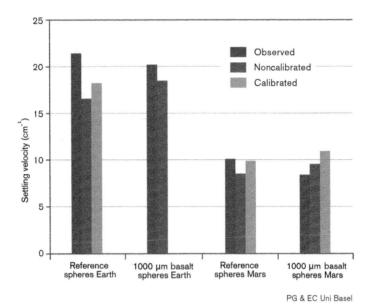

Fig. 11.2. Observed and calculated settling velocities for reference and 1000 μm basalt spheres used for MarsSedEx II. No bar is shown for the data set used for calibration.

but both by a fairly small (9% and 13%) margin. Calibrating the combined model for basalt spheres on Earth is done by adjusting C_1 because C_2 is supposed to describe particle shape and is set at 0.4 for spherical particles. A good fit on Earth for the basalt spheres is achieved by reducing C_1 from 18 to 10.3. Lowering C_1 for Earth indicates that the basalt spheres are hydraulically smoother than the particles used for the development of the combined model. Calibrating C_1 for the basalt spheres still underestimates the settling velocity of the reference spheres on Earth (−15%), confirming their lower skin friction. For Mars, the calibration has the opposite effect: the reference spheres are predicted almost correctly (+3%), but the basalt spheres are overpredicted by 30%. The poor fit of the calibrated model for the basalt spheres on Mars is plausible because the relevance of viscous drag is probably underestimated by a C_1 fitting for Earth. The good fit for the reference spheres just reflects a coincidental correction in the right direction, i.e., from a far too high description of viscous drag to a lower one.

11.5 DIAMETER AND SETTLING VELOCITY OF BASALT SPHERES

The basalt spheres ranging nominally from 500 to 1000 μm diameter have the density of a silicate, so a good fit with the noncalibrated reference model was, as in the previous section, expected for this part of the experiment. The quality of the noncalibrated combined model and the calibration developed in the previous section for the 1000 μm basalt spheres is used again to test their applicability to the settling velocities of smaller spheres at reduced gravity.

Figure 11.3 shows the observed and model results for terrestrial and simulated Martian gravities. The observed increase in settling velocity with size is plausible and follows the change in diameter. The noncalibrated calculated values for terrestrial gravity were all smaller than observed settling velocities. The error increased with declining particle size. This confirms that the drag of all the basalt spheres is smaller than the coefficients derived during the development of the combined model

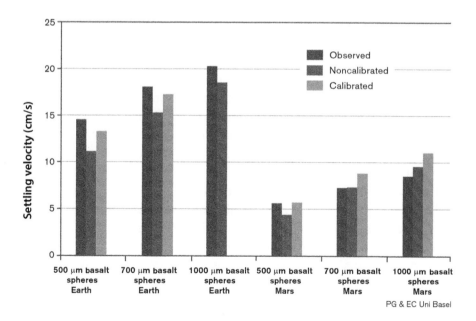

PG & EC Uni Basel

Fig. 11.3. Observed and calculated settling velocities for basalt spheres of 500, 700, and 1000 μm diameter used for MarsSedEx II. No bar is shown for the data set used for calibration.

and reported in the corresponding literature suggest. Assuming that the combined model represents the spherical shape with a $C_2 = 0.4$ well, all of the basalt spheres would therefore have a smaller viscous drag. The increasing model error with declining particle diameter supports this assumption because the relative effect of viscous drag on terminal velocity is greater for small particles.

For Mars, the quality of the prediction was, somewhat surprisingly, better than for Earth. Especially the error for the 0.7-mm sphere was almost negligible (-0.03 cm/s or 0.4%). Being the medium-sized sphere, one would expect that these parameter values would generate also a good fit for the other particles. However, the calculated settling velocity of the 1000 μm spheres was too high while the model prediction for the 500 μm was too low. These errors indicate that the noncalibrated combined model underrates the increasing turbulent drag with increasing particle size for Martian gravity, leading to model results that are too high for the 1000-μm spheres. For small particles, the difference between observed and predicted settling velocity illustrates the opposite effect: the model assumes a greater drag and thus lower settling velocities, which indicates that the increasing relevance of viscous drag is captured by C_1, but still overrated by using the value fitted for Earth.

Using the C_1 calibrations developed for the 1000 μm basalt spheres on Earth in the previous section reduces the underprediction of settling velocities on Earth to 10% or less. For Mars, the reduced C_1 works better for the 500-μm spheres (error = 0.08 cm/s or 1.5%) than the 30% error seen in the previous section for the 1000-μm spheres. The contrast between the good fit for the large and small spheres indicates that the increasing effect of turbulent drag on Mars is not reflected well by the calibrated combined model, confirming the pattern of errors generated by the noncalibrated model. We attribute this problem to the change in the ratio of viscous versus turbulent drag, or the transition between Stokes and Newtonian flow, between settling on Earth and Mars. The combined model does apparently not capture the exact pattern of this shift properly. Clearly, a simple "fix" of the combined model to reflect the transition zone better is not possible without a greater database to develop a more comprehensive solution.

11.6 PARTICLE SHAPE EFFECTS ON SETTLING VELOCITY

Comparing reference and basalt spheres in the previous sections illustrated that the combined model requires calibration for particles of different size, even when, such as for the basalt spheres, consisting of the same material and having the same shape. Moving toward natural particles, the next test focuses on the effect of shape on settling velocity and the suitability of calibration on Earth to capture the effect on Mars. For this part of the MarsSedEx II tests, four particles of roughly the same diameter and settling velocity, but with increasing deviation from a spherical shape, were selected (Table 10.1). The hematite had also a density of 4.3 g/cm^3, much greater than the other three types of sediment that were tested, which ranged from 2.5 to 2.77 g/cm^3. To give an indication of the lack of the declining roundness, the specific surface area of the particles is given in Table 10.1. The poor visibility of the Amazon sand particles on the video clip recorded during the reduced gravity flight led to a change of the size used for modeling from 1 to 1.4 mm. This correction was based on a comparison of the number of particles that were clearly visible with those of other size classes. The difference indicated that only the largest Amazon sand particles could be counted properly. Therefore, the D_{90} instead of the D_{50} was used for modeling. Accordingly, the results referring to the Amazon sand require some cautious interpretation.

The observed settling velocities on Earth were smaller than expected based on the sediment properties amounting to an overall 25% difference between the fastest settling hematite and the slowest, the basalt sand. This is smaller than the differences in particle properties suggest. Unlike the basalt spheres, the errors between the noncalibrated model and the observation are also fairly small (<10% error) for all tested particles, implying that the combined model captures the effect of the varying sediment properties on settling velocities well within a small range of settling velocities.

The estimated velocities were, except for the basalt spheres, greater than the observed. The lower viscous drag of the basalt spheres identified in the previous sections is probably the reason for this error. The overprediction for the other particles is plausible because they are rougher than spheres, which requires an increase of the C_2 value of 0.4 to reflect the greater turbulent drag.

For simulated Martian gravity, the differences between the observations and the noncalibrated combined model were fairly varied. Settling velocities of the Amazon sand were almost correctly predicted by the model while the basalt sand was overpredicted by up to 30%. For the hematite, on the other hand, the model result was too low (14%). The overprediction for Amazon and basalt sands can be explained like for Earth because the default C_2 value does not properly reflect particle roughness. Depending on the change of slope between Reynolds number and drag coefficient (Figure 4.3), the effect of increasing form roughness might be stronger at low settling velocities, which would explain the greater error for the basalt compared to the Amazon sand. For the hematite, the underprediction is less plausible. A possible explanation might be its very blocky shape, which may lead to an orientation of particles in the water along their longest axis, thereby effectively reducing the diameter causing friction. This effect might be more pronounced for Mars than Earth as the video clips suggest. On Earth, at greater settling velocities, many hematite particles wobbled along their downward settling path. Such a movement was observed to a lesser degree on Mars, suggesting smoother flow with less drag. But this explanation has to remain a speculation waiting for further testing.

Adjusting the C_2 for each particle on Earth (Table 11.3) and applying these values for simulated Martian gravity delivers a further indication about the relevance of flattening turbulent drag at greater settling velocities for the tested particles. Using the C_2 values for the combined model calibrated for Earth to predict Martian gravity settling velocity reduces

Table 11.3. C_1 and C_2 Coefficients Generating Best Fits for MarsSedEx II Model Calibration		
Sample	C_1	C_2
500 μm basalt sphere	10.3	
700 μm basalt sphere	18	
1000 μm basalt sphere		0.32
Amazon sand		0.485
Hematite sand		0.455
Slow settling basalt sand		0.635
Medium settling basalt sand		0.58
Fast settling basalt sand		0.5

the overprediction for the basalt sand, but generates a greater underprediction for both the Amazon and hematite sands than the noncalibrated model. Apparently, on Mars turbulent drag caused by shape is greater for particles in the observed size range than the calibration for Earth suggests. The pronounced "edginess" of the hematite particles, combined with their high density, would induce more turbulent drag than the rounder-shaped basalt and Amazon sands experience. This effect might be overrated by a C_2 value fitted for Earth at lower settling velocities on Mars. Overall, the different qualities of the noncalibrated and calibrated model outputs for Martian and terrestrial gravities illustrate the limitations of the combined model to capture drag in the transition zone between Stokes and Newtonian flow when using a calibration from Earth for Mars.

11.7 SETTLING OF "REAL" MARTIAN SEDIMENT

The final set of sediment particles used during MarsSedEx II aimed at testing the quality of the combined model for predicting the settling velocity of basalt sand that resembles sediment on Mars. Due to the limitations of particle properties used to describe sediment size (see Figure 4.7) to capture the effect of shape on drag and thus, their settling velocity, the sediment was not chosen by size, but by settling velocity. Three ranges of settling velocities, 9.5–12, 12–16, and 16–25 cm/s, corresponding to noncalibrated combined model results for spheres of 500, 700, and 1000 μm and a density of 2.65 were chosen. To achieve a good fractionation of the sample, the separation was carried out using the 1.8-m Basel settling tube described in Chapter 5. The individual character of the basalt sand particles in each of these three size classes, and thus the relevance of shape and density for sediment formed from a multimineral rock, is illustrated by the contrast of very small differences in D_{50} and specific surface area, while density increases with settling velocity. It is still noteworthy that these differences are smaller than required to generate the settling velocity difference, which illustrates our generally limited understanding of particle–water interaction (Figure 11.4).

The differences of particle properties between settling velocity classes translate into the settling velocities observed using the MarsSedEx II instrument on Earth (Figure 11.5). The settling velocities still fall into the ranges of the fractionation except for the smallest particles. Their slightly greater settling velocities are caused by the lower temperatures

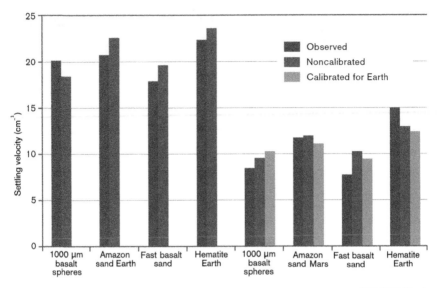

PG & EC Uni Basel

Fig. 11.4. Effect of roundness on settling velocities for a basalt sphere, and a quartz, basalt, and hematite sand of declining roundness.

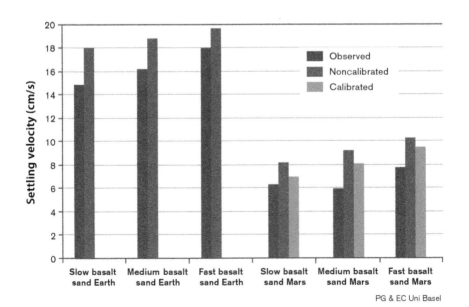

PG & EC Uni Basel

Fig. 11.5. Observed and estimated settling velocities for a basalt sand ranging in settling velocities between 9 and 25 cm/s².

of the water used during the fractionation than during the flight and ground tests. For simulated Martian gravity, settling velocity also increases slightly except for the medium basalt sand. This is attributed to the lower simulated gravity compared to the fast settling basalt sand during the portion of the parabola when the measurement took place (Table 11.1). While still greater than for the slow settling basalt sand, the low observed settling velocity of the medium and fast basalt sand also hint at a strong effect of increasing settling velocity on drag.

Comparing observed settling velocities and noncalibrated model output for Earth reveals an overprediction, which can be explained by the lack of consideration given to the nonspherical shape of the sediment by setting C_2 to 0.4. The error is declining for particles with a greater original settling velocity, indicating that the relationship between Reynolds number and drag moves into a region where it begins to flatten with increasing particle size on Earth (see Figures 4.5 and 4.7). For Mars, the noncalibrated model also overpredicts all settling velocities, but with a greater proportional error, up to 35% for the medium sand, than for Earth. This varying fit of the noncalibrated model for both Earth and Mars confirms that the transition zone between Stokes and Newtonian flow is not captured well by the combined model.

Calibrating the combined model by adjusting C_2 (Table 11.3) to account for the nonspherical shapes generates best fits for values decreasing from 0.635 to 0.5 on Earth. The adjusted C_2 values show that the relative importance of turbulent drag is declining with settling velocity of the particles. This confirms the observation made above that the relationship between Reynolds number and drag coefficient shifts toward a flatter slope (see also Figures 4.5 and 4.7). It also shows the need for adjusting coefficients to size or settling velocity of particles, rather than using one C_2 parameter value for one type of sediment.

Using the calibration of C_2 values for Earth for the simulated Martian gravity leads to an improved fit for the fast settling, particles, but poorer ones for the medium- and fast-settling sands. This indicates that the ratio between viscous and turbulent drag is apparently captured well for slow-settling sediment, but does not work well for faster-settling particles anymore. This confirms that the sensitivity of turbulent drag to changing settling velocity is still more pronounced on Mars than for the same particle settling on Earth. The different model output quality for C_2 values fitted on Earth for Mars thereby confirms the problem identified in Chapter 4: the shift from Stokes to Newtonian flow generates a change of drag. Therefore, drag coefficients generated on Earth cannot be used on Mars, at least for particles of a critical size and settling velocity in a range between the dominance of either flow regime. So far, it is also unclear when exactly the transition occurs for a given particle.

11.8 SUMMARY OF MarsSedEx II OBSERVATIONS AND MODEL RESULTS

The analysis of the MarsSedEx II data based on comparing the observed values with a noncalibrated and a calibrated model for Earth shows that the combined model produces a fair quality of settling velocities. Generally, a good fit (either noncalibrated or calibrated) for one type of sediment on Earth does not generate a similarly good fit on Mars and vice versa. When a good fit is achieved for both gravities, it does not extend to particles that differ with regards to another property that is supposed to be captured by the combined model. The errors for the basalt spheres illustrate this observation: adjusting C_1 for 1000 μm spheres on Earth caused an overprediction for Mars, i.e., limited recognition of viscous drag. On the other hand, a good fit on Mars, for example, for the 500 and 700 μm particles, was associated with an overprediction on Earth. The effects of particle shape were captured reasonably well for the large and still fairly spherical Amazon sand, but generated both too high and too low values for the other particles. The roundness comparison conducted with sediment of approximately 1-mm diameter shows that particles of this size appear to be already in the realm of a flattening relationship between turbulent drag and settling velocity. Consequently, poorer fits using coefficients from Earth can be expected for smaller particles, as is illustrated by the basalt sand where calibration does not work well. Overall, the results indicate that the use of the combined model requires a recalibration, best according to particle size, for Mars.

A further interesting result of the MarsSedEx II mission is the distribution of all observed settling velocities for the basalt sand (Figure 11.6). On Mars, the range is much smaller than on Earth. This confirms one of the tentative results of MarsSedEx I: sorting from a given layer of moving, shallow, water is less pronounced for

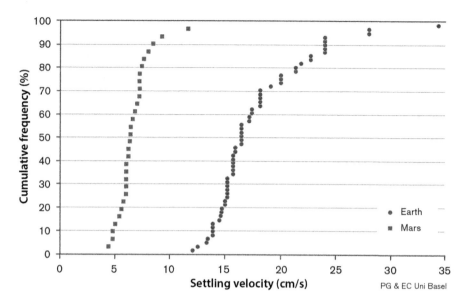

Fig. 11.6. Cumulative frequency distribution of observed settling velocities for basalt sand during MarsSedEx II mission.

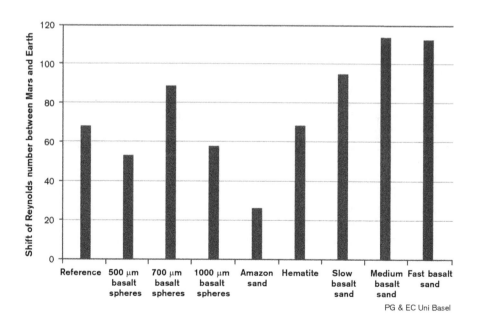

Fig. 11.7. Increase of particle Reynolds number between Mars and Earth, based on observed settling velocities. The shifts are small compared to the range of particle Reynolds numbers observed under natural conditions, but occur in the sensitive transition between Stokes and Newtonian flow. It is also noteworthy that they differ significantly, which demonstrates that the particles are affected not uniformly by the gravity-induced change of flow hydraulics.

Mars than for Earth. This is attributed to the shift in flow hydraulics, which give smaller emphasis to particle size for determining drag in the tested sand range on Mars than on Earth. This transition of flow conditions is also illustrated by the shifts in particle Reynolds numbers for the observed settling velocities between Martian and terrestrial gravities (Figure 11.7). Depending on the particle, the shift is more or less pronounced. This confirms that different types of sediment particles vary in their sensitivity to settling "similar to Earth" on Mars. This conclusion confirms the theoretical considerations of Chapter 4. The relevance for Mars will be illustrated and discussed in the final chapter.

BIBLIOGRAPHY

Ferguson, R.I., Church, M., 2004. A simple universal equation for grain settling velocity. J. Sedimentary Res. 74, 5.

Grotzinger, J.P., Hayes, A.G., Lamb, M.P., McLennan, S.M., 2013. Sedimentary Processes on Earth, Mars, Titan, and Venus. In: Mackwell, S.J. et al. (Ed.), Comparative Climatology of Terrestrial Planets. University of Arizona, Tucson, pp. 439–472.

Grotzinger, J.P., Sumner, D.Y., Kah, L.C., Stack, K., Gupta, S., Edgar, L., Rubin, D., Lewis, K., Schieber, J., Mangold, N., Milliken, R., Conrad, P.G., DesMarais, D., Farmer, J., Siebach, K., Calef, F., Hurowitz, J., McLennan, S.M., Ming, D., Vaniman, D., Crisp, J., Vasavada, A., Edgett, K.S., Malin, M., Blake, D., Gellert, R., Mahaffy, P., Wiens, R.C., Maurice, S., Grant, J.A., Wilson, S., Anderson, R.C., Beegle, L., Arvidson, R., Hallet, B., Sletten, R.S., Rice, M., Bell, J., Griffes, J., Ehlmann, B., Anderson, R.B., Bristow, T.F., Dietrich, W.E., Dromart, G., Eigenbrode, J., Fraeman, A., Hardgrove, C., Herkenhoff, K., Jandura, L., Kocurek, G., Lee, S., Leshin, L.A., Leveille, R., Limonadi, D., Maki, J., McCloskey, S., Meyer, M., Minitti, M., Newsom, H., Oehler, D., Okon, A., Palucis, M., Parker, T., Rowland, S., Schmidt, M., Squyres, S., Steele, A., Stolper, E., Summons, R., Treiman, A., Williams, R., Yingst, A., Team, M.S., 2014. A habitable fluvio-lacustrine environment at Yellowknife Bay, Gale Crater. Mars. Sci. 343, 14.

Julien, P.Y., 2010. Erosion and Sedimentation, Second Edition. Cambridge University Press, Cambridge, p. 371.

Knighton, D., 2014. Fluvial Forms and Processes: A New Perspective, Second Edition. Routledge, New York, p. 383.

Komar, P.D., 1979. Modes of sediment transport in channelized water flows with ramifications to the erosion of the Martian outflow channels. Icarus 42, 13.

Melosh, H.J., 2011. Planetary Surface Processes. Cambridge University Press, Cambridge, UK, p. 500.

CHAPTER 12

Outlook: More Experiments or Better Models for Sedimentation on Mars?

ABSTRACT

The successful completion of the Mars Sedimentation Experiment missions documented the problems associated with the use of semiempirical models for sediment settling velocities developed for use on Earth when applying them to the reduced gravity on Mars without further calibration. In this chapter, the specific implications of using noncalibrated models for the identification of potential traces of life-bearing strata are identified. Moving on, the risks induced by ignoring gravity associated with other models describing surface processes, such as flow hydraulics and sediment transport, are discussed. Concluding this chapter and the theme of the book, the role of further experiments aimed at supporting the ongoing exploration of Mars, e.g., during the upcoming ExoMars and InSight missions, is presented.

12.1 MarsSedEx AND SURFACE PROCESSES ON MARS

In this final chapter the implications of MarsSedEx I and II for current Mars exploration activities, including upcoming missions, are discussed and the future direction of experimental research on surface processes on Mars is sketched out. The MarsSedEx I and II results presented in Chapters 9 and 11 show that both missions fully achieved their aims and objectives. The two missions demonstrated that sediment settling on Mars can be studied using settling tubes during reduced gravity flights.

Experiments in Reduced Gravity: Sediment Settling on Mars. DOI: 10.1016/B978-0-12-799965-4.00012-1

They also confirmed, as speculated in Chapter 4, that the effect of gravity on settling velocity changes the flow hydraulics around a particle moving in water so significantly that semiempirical models used in Earth cannot be used on Mars without calibration, unless the risk of an error on the order of 30–50% is accepted. In addition to identifying the shortcomings of empirical models for a precise description of particle settling on Mars, the observed settling velocity data also reveal marked differences in the range of settling velocities between Earth and simulated Martian gravities for a sediment resembling sands on Mars (Figure 11.6). Most notably, the range of settling velocities for reduced gravity is much smaller than for Earth. This is of critical importance for assessing sedimentation processes on Mars, and thus, habitability and/or likelihood of preserving traces of life. This smaller range of settling velocities implies that in a layer of water of a given depth and velocity, sediments on Mars would be less sorted than on Earth.

Figure 12.1 illustrates the implications of less pronounced sorting for the assessment of depositional environments on Mars. The high-resolution image taken by Curiosity's Mars Hand Lens Imager (MAH-LI) on sol 400 of the mission (September 21, 2013) shows a conglomerate rock called *Darwin* in Gale crater. The texture of the particles is not uniform and faint strata of varying frequency of particles >1 mm are visible. The size distribution of large grains, based on a count of 120 clearly identifiable sediment particles ranging from 0.4 to 14 mm in diameter, is shown in Figure 12.2. The coefficient of variation, skewness, and kurtosis can be used to assess the sorting of the measured particles. Coefficient of variation (1.5) and kurtosis (10.0), indicate a rather poor sorting while the skewness (3.03) toward the large particles shows that they are sorted better than the smaller ones. Including the noncounted matrix in the particle size distribution would presumably lead to a more balanced skewness, but widen the coefficient of variation and kurtosis.

One can only speculate how the same sediment would be sorted on Earth because the controlling factors for flow depth, velocity, and turbulence are not independent of each other. Still, at least the qualitative effect of limited sorting on sediment properties can be assessed. The second frequency distribution in Figure 12.2 shows a floodplain deposit from the Cuckmere river in South East England. While probably not a

Fig. 12.1. Darwin conglomerate rock in Gale crater, taken by Curiosity's MAHLI on sol 400 of the mission. NASA/JPL Photojournal PIA17362.

suitable analog for river deposits on Mars, the size distribution of this sediment represents an environment with a high potential for habitability because of the proximity to permanently flowing water. It is also a depositional environment with a great potential to trap and preserve organics, i.e., traces of life. The more pronounced peak, small kurtosis (0.125) and coefficient of variation (0.51) of the Cuckmere distribution illustrate a much better sorting. Based on our settling velocity data, one could speculate that the Darwin conglomerate would look a bit more like the Cuckmere if deposited on Earth, having lost the large particles before the fines were deposited in the floodplain because of the more pronounced potential for sorting at terrestrial gravity.

The potential differences in sorting between Mars and Earth highlight a problem when looking for sites with evidence for past habitability

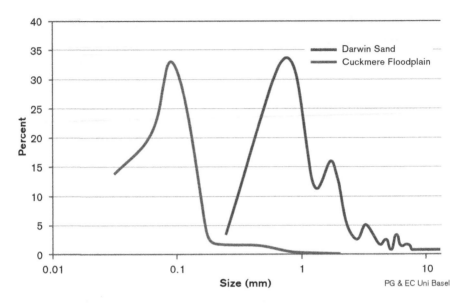

Fig. 12.2. *Frequency distribution of sand-sized particles of the Darwin conglomerate, Gale crater, Mars, and the Cuckmere river floodplain, Sussex Downs, England, Earth. Data for the Cuckmere floodplain.* Courtesy of Philip Greenwood, University of Basel, Switzerland.

on Mars. On Earth, depositional environments rich in fine, potentially organics preserving sediments, form in two domains: first, slopes of 1.5°–3° at the base of alluvial fans, and second, in floodplains next to rivers with slopes of 0.5°. These two environments differ in their potential habitability because alluvial fans are wet for short periods of time only, while floodplains may experience a prolonged inundation. Due to the lower gravity on Mars, the flow hydraulics associated with these depositional environments are likely to form already at greater slope angles. A precise estimate of the effect, however, is limited by the empirical nature of the models used to describe flow hydraulics and sediment transport. Combined with the poorer sorting at Martian gravity observed during MarsSedEx I and II, this has implications for the selection of strata to search for traces of life. On Earth, sediment similar to the one forming *Darwin* is likely to be found either on low sloping alluvial fans with short flushes of runoff that ceases suddenly and thus limits sorting, or in a streambed experiencing similar short flushes of runoff. On Mars, on the other hand, hydraulic conditions leading to deposition from permanent or at least seasonal or prolonged perennial channelized flow would experience deposition at steeper slope angles than on Earth. Due to the limited differentiation of settling velocities, at least in shallow flow, the sorting of these sediments would be less pronounced than on Earth. As a consequence, one can speculate, that fluvial sediment with traces of life might look more like a fan deposit on Earth considered to be less habitable than a floodplain. The need to distinguish the different sedimentary environments, including the origin of the sediment, as precisely as possible is illustrated by the proximity of the strata at *Yellowknife Bay* and their apparent lack of organics.

12.2 MODELING FLUVIAL PROCESSES ON MARS

The smaller range of settling velocities observed for simulated Martian gravity during MarsSedEx II are indicative of less pronounced sorting of sediment on Mars than on Earth from similar flows. Arguably, the reduced gravity on Mars is also likely to reduce flow velocities on a given slope and thus, for a given volume of water moving downslope, increase water depth. Both slower flow velocities as well as deeper flow would counteract the effect of a smaller range of settling velocities on sorting. But none of these effects are linear and the numerical models used to calculate them suffer from problems similar to those identified when using settling velocity models developed for Earth on Mars. Putting the MarsSedEx I and II results into the context of identifying habitability of Martian depositional environments and potential to preserve traces of life highlights the need for linking sedimentation properly to the flow hydraulics. This requires testing and where required the adjustment of models used on Earth to achieve the precision required for describing and distinguishing target sediments for study by landers and rovers from less interesting ones. Apart from the settling of particles examined in this book, the description of further processes such as the flow hydraulics of running water and the entrainment and movement of sediment particles has to reflect the effect of reduced gravity.

In Chapter 1, the critical issues of models used to describe flow velocity and the critical shear stress required to move sediment were identified. Concluding this reflection on MarsSedEx I and II, the common approaches are examined again more critically in the light of the results of the two missions. Flow velocity depends on the force pulling the water downslope and the friction of the liquid between surface and moving water. The force pulling water downslope depends on the mass of water, gravity, and the slope angle.

$$F_g = \sin(\alpha)mg$$
Equation 12.1

where

α = slope angle
m = mass of water
g = gravity

For Mars, this force is reduced by the proportional difference of gravity between Earth and Mars, i.e., approximately 62%. The roughness of the streambed and the depth of the water layer affected by the roughness elements determine the friction between the water moving downslope. This generates a water depth-dependent flow resistance. To avoid using functions describing this change for a given surface depending on water depth, roughness factors have been developed that can be used to characterize the hydraulic roughness of a streambed by a single parameter value. The Manning coefficient described in Chapter 1 illustrates this approach and highlights the relevant factors by estimating flow velocity as

$$v = \frac{k_n}{nR^{2/3}S^{1/2}}$$
Equation 12.2

where

v = cross-sectional average velocity (m/s)
k_n = 1.0 for SI units
n = Manning coefficient of roughness

The results of MarsSedEx I and II indicate that applying this equation on Mars generates a problem because the exponents for hydraulic radius and slope have largely been derived by fitting equations to observations on Earth. The Manning coefficient can be seen as a factor describing the drag of a streambed. On Mars, slower runoff might therefore experience a smaller friction between bed and water, as was seen for the faster than predicted settling particles at reduced gravity.

Apart from the value of the Manning coefficient, the equation above also modifies slope and hydraulic radius by empirically fitted coefficients. This approach is not problematic under conditions where runoff occurs in a certain domain with fixed boundary conditions, such as gravity, but exponents from Earth are unlikely to work for different planets because they do not fully reflect the way the physical processes and controlling surface and sediments interact. Selecting models for application on Mars should therefore avoid as much empiricism as possible. Using a different approach to quantify streamed friction, the Darcy–Weisbach friction factor, illustrates this approach. The Darcy–Weisbach friction factor is expressed by

$$ ff = \frac{8gds}{v2} \qquad \text{Equation 12.3} $$

where

ff = Darcy–Weisbach friction factor
g = gravity
d = flow depth
s = bed slope
v = flow velocity

The information required to calculate the Darcy–Weisbach friction factor has to be collected in the field. The value of the friction factor for a given streambed is, like for settling particles, determined by flow hydraulics and thus, closely related to the Reynolds number (Figure 12.3). As for settling velocities, the relationship varies with Reynolds number, e.g., *ff* = 64/Re for Reynolds numbers below 2000, but flattens for greater values. Since the Darcy–Weisbach friction factor basically describes the drag of a streambed on a moving layer of water, applying it to Mars potentially generates the same problems than those identified for drag coefficients of settling particles in this study. This leaves a good estimate of flow velocities using simple, empirically derived equations working on Earth at least in need of a calibration for Mars.

The problem of the applicability of semiempirical models developed for Earth on Mars extends to the assessment of sediment entrainment, transport, and deposition. The critical flow property determining which particles can be moved is the boundary shear

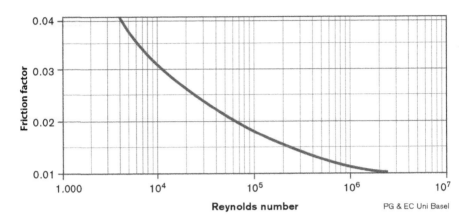

Fig. 12.3. Schematic illustration of the relationship between Reynolds number and streambed friction factor. Note the resemblance of the curve to Reynolds number and drag coefficient.

stress τ of the water acting on the surface. The boundary shear stress can be expressed as

$$\tau = \rho g d S \qquad \text{Equation 12.4}$$

where

ρ = density of the fluid
g = gravity
d = flow depth
S = slope of flow

The entrainment of sediment particles is associated with a critical shear stress, called the Shields parameter after the German engineer A.F. Shields who developed this method in the 1930s. The critical shear stress is calculated by relating the boundary shear stress to the force of gravity acting on the sediment particle and written as

$$\tau^* = \Theta = \frac{\tau}{(\rho_s - \rho)gD} \qquad \text{Equation 12.5}$$

where

$\tau^* = \Theta$ = Shields parameter for a given sediment particle
τ = boundary shear stress [Pa]
ρ_s, ρ = density of the sediment and the fluid [g/cm^3]

g = gravity [m/s²]
D = particle diameter

The Shields parameter is not independent of flow conditions, generally declining for a given particle with an increase of turbulence, i.e., another relationship between a coefficient linking an observed sediment reaction to driving forces based on the Reynolds number!

For Mars, the limitation of using Shields parameter values developed for Earth has two potential problems. First, the calculation of the boundary shear stress is based on flow depth, which for a given volume of water moving downslope requires information on flow velocities. In the light of the shift in drag coefficients observed during MarsSedEx I and II, the fit of friction factors estimated for Mars based on empirical relationships for Earth should at least be tested, i.e., in flumes carried onboard reduced gravity flights. The second problem relates to the friction between a particle and the surface below (Figure 12.4), which affects the value of the Shields parameter. Friction between fixed beds and loose material results from the complex interaction of the force of gravity pulling a particle onto the surface and the adhesion and interlocking of particles on the streambed. As for settling and flow velocity, each of

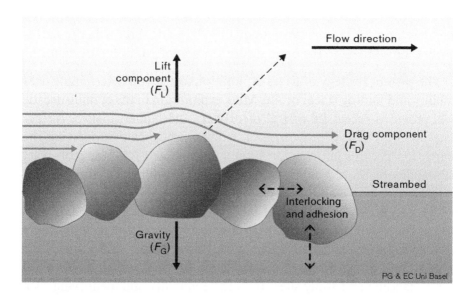

Fig. 12.4. Forces acting on a particle on the streambed.

these factors is not independent of gravity, either directly or indirectly, and lumped into the Shields parameter. Therefore, like settling and flow velocity, the applicability of the empirical relationships used on Earth needs to be tested for Mars. This conclusion is supported by results of the experiments done by Maarten Kleinhans and coworkers mentioned in Chapter 1, who found that the angle of repose differs for Martian gravity, showing that assuming independence of friction between loose particles and gravity is not the case.

12.3 MOVING FORWARD: FOCUS ON MARTIAN HYDROLOGY

The MarsSedEx research was motivated by a critical review of the applicability of semiempirical models used on Earth to describe the movement of particles in water on Mars. The initial part of the research focused on sediment settling because it relates to a number of other processes relevant to flow hydraulics and sedimentation. To achieve this aim, instruments that could measure settling velocity in the constrained conditions of a reduced gravity flight had to be developed and tested. The results show that experiments on sedimentation are possible in reduced gravity and that common, empirically derived models for Earth are actually limited in their use on Mars. The reduced gravity flights also showed that both the instruments as well as the range of sediments tested require further development to improve data quality. To guide this further research, a set of aims linked to the ongoing exploration of Mars can be defined.

The current strategy of Mars exploration focuses on determining habitability and finding traces of life. Mars exploration in the coming decade will, provided successful arrival of orbiters and landers, deliver further data on the physical properties of sedimentary rocks. Supporting the search for life, this information should be exploited fully to reconstruct habitability of Mars. Understanding the past hydrology of Mars is a key to assessing its habitability. Rock-forming sediment records the environmental conditions at the time of its deposition. Unlocking the information in these archives, based on a precise understanding of the differences between sedimentation on Earth and Mars, will greatly contribute to our understanding of the environmental history of Mars and guide the selection of landing and study sites. Tracing habitability and identifying rock formations most likely to contain traces of life (if it existed on Mars) require to determine the presence of liquid on Mars more precisely than currently possible, i.e., the regime of runoff in rivers, such as the likelihood of perennial or episodic runoff. In addition, our knowledge of the changing geographic distribution of water during the Mars past has to be improved. Achieving a good understanding of Martian paleohydrology will enable a more targeted search for past habitats and traces of life. Once on the ground, precise models on sedimentation will provide a tool to select the most promising sites for the analysis of surface rocks by supporting the assessment of the depositional environment they were formed in. This extends to identifying coring sites with the most promising stratigraphy to contain and preserve traces of life (Figure 12.5).

The search for past habitats and environmental archives containing traces of life requires combining imagery taken from orbit with models of surface processes, as was done, for example, for the landing site selections of Curiosity and currently ExoMars. First, geomorphic evidence for runoff and deposition in the past is identified, followed by modeling of hydrology, erosion, and sedimentation to identify the most likely location of potential past habitats and/or archives carrying traces of life. Using an analog, such modeling can be seen as a tool to find a vanished

Fig. 12.5.   Cumberland rock with drilling hole and depression created by ChemCam laser. Image taken by Curiosity's Mast Camera on sol 281 of the mission (May 21, 2013). Understanding sedimentation on Mars can support the selection of the most promising sites to look for traces of past life on Mars by drilling into the rocks most likely to have contained and/or preserved these traces. Credit: NASA/JPL-Caltech/MSSS PIA 17069.

oasis in a desert by generating precise information on the spatial pattern, magnitude, and frequency of runoff, erosion, and deposition. Calibrating the numerical models of surface processes, among others the ones introduced in this book, will therefore strongly support the process of landing and/or target site selection for the search for habitability and traces of life by rovers and other landers.

12.4 LINKING MODELS FROM EARTH TO MARS

The potential shortcoming of surface process models on Mars and the need to analyze images of rocks on Mars, both to study environmental history as well as to support the selection of drilling sites, leads to the question of how the knowledge gaps identified in this book can be closed. The feasibility of developing instruments for the calibration of existing models using reduced gravity flights has been shown by MarsSedEx I and II. Continuing such experiments with more particles, especially realistic mixtures of sediment as well as water properties, i.e., Martian brines, can contribute to improving model outputs. The development of instruments to observe flow hydraulics such as small flumes is also possible. However, there are limitations set by both the constraints of reduced gravity flights as well as the wide range of conditions encountered on Mars. During a rover or lander mission when quick decisions on research targets are required, a timely experimental approach testing a wide range of options is not feasible. Consequently, there is a need for the development of more elaborate models to simulate the relevant surface processes.

A theoretical alternative to recalibrating empirical models for Mars is the use of computational fluid dynamics (CFD) models. However, applying the basic principles of physics to complex problems such as sediment movement is restricted by the state of knowledge of the boundary conditions required to run these models. For sedimentation, this would, for example, include a sufficient description of particle shape, which is possible for single particles, but limited by the wide variety of naturally occurring ones and our ability to define relevant properties, e.g., roughness determining skin friction, on Mars. CFD, on the other hand, can be used to run sensitivity tests, for example, by comparing and adjusting CFD models to observe settling velocities, followed by using a well-working model to test sensitivity of settling velocity to an increasing complexity of shapes. These studies can be extended to issues such as hindered settling of particles from suspension, the values of the Shields parameter for reduced gravity and the friction between surface and running water. A good set of such models can also be used to make a quick assessment of the depositional environment during the exploration of the Martian surface by a rover and thereby support the decision on where samples should be taken or cores should be drilled.

Combining reduced gravity experiments with the development of CFD models is limited by the scale, both in space and time, of instruments and the measurements that can be conducted during a reduced gravity flight. For example, any deposition process requiring more than 25 s to complete cannot be studied. Similarly, analog experiments used to calibrate or validate numerical models on landform development, such as the depositional fans shown in Figure 12.6, are too large to put onboard a plane and probably also not safe to fly. Therefore, a second

Fig. 12.6. Analog model of fan development on a 9 m² plot at the University of Exeter, UK. Photo courtesy of Lucy Clarke, British Antarctic Survey, Cambridge, UK.

link to smaller scale experiments should be developed by testing analog setups on Earth, e.g., by identifying a sediment with lower density which behaves on Earth similar to natural mineral grains under reduced gravity. Such Mars-analog sediment could then be used to study larger-scale processes over longer periods of time.

Finally, some processes may need further basic research first before assessing their relevance on Mars. One of these is the settling in brines. In saltwater, suspended particles tend to coagulate on Earth leading to much increased settling velocities. Even on Earth the process is not fully understood or quantified and requires basic research first. However, research in reduced or even prolonged zero gravity on coagulation in brines might be useful because by eliminating the settling, the time for coagulation is extended, and not disturbed by the motion through water. A similar research gap extends to the settling and sorting of mixtures of mineral and organic particles. Finally, the structure of the sedimentary rock encountered in an outcrop depends on the weight of the overburden at the time of lithification. This introduces gravity and raises the question whether, apart from sorting, the taphonomy of target strata differs between Earth and Mars.

The review of the shortcomings of the empirical models used to simulate surface processes and landform development on Mars in this book illustrates the need for further experimental research. Based on the identification of knowledge gaps highlighted above, such experimental research should follow a strategy that focuses on increasing and supporting the scientific output of current and future rover and lander missions to Mars. Three aims for the research using experiments can be set. First, a reliable and quick assessment of the nature of the depositional environment of a landing site should be achieved; this involves both the ability to interpret the sediment strata as an environmental archive as well as their likelihood to bear traces of past life. Second, the ability of modeling the development of larger-scale landforms aimed at the identification of the constraining climatic conditions and thus, paleohydrology should be improved to gain a better understanding of the environmental history of Mars and the temporal and spatial dynamics of water at its surface. Finally, the plausibility of the association of changes in surface properties

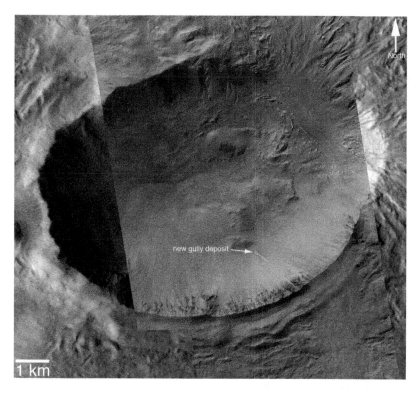

new gully deposit

1 km

Fig. 12.7. A gully forming from groundwater seepage today? Mars Global Surveyor image with a light toned gully feature in the center of the wall of an unnamed crater in the Centauri Montes Region. Credit: NASA/JPL/Malin Space Science Systems PIA09028.

observed today and attributed to running water, should be tested (Figure 12.7).

These three strategic goals apply to a certain extent to Earth as well. Improving our ability to simulate the formation of sedimentary rocks and landforms on Mars will contribute to identify the reason for the different developments the two planets took and thus, ultimately assess the stability of habitable conditions on Earth. More practically, improving models for surface processes will enable the identification of risks associated with global environmental change on Earth, for example, the pollution of waters, and thus also contribute to, if not habitability, but at least quality of life on Earth. To conclude, there is merit (and merriment) in the experiment!

BIBLIOGRAPHY

Anderson, R.B., Edgar, L., Bridges, J.C., Williams, A., Williams, J., Ollila, A., Forni, O., Mangold, N., Lanza, N., Sautter, V., Gupta, S., Blaney, D., Clark, B., Clegg, G., Dromart, G., Gasnault, O., Lasue, J., Le Mouélic, S., Leveille, R., Lewin, E., Lewis, K., Maurice, S., Nachon, M., Newsom, H., Vaniman, D., Wiens, R.C., 2014. ChemCam results from the Shaler Outcrop in Gale Crater, Mars. 45th Lunar and Planetary Science Conference.

Arvidson, R.E., Squyres, S.W., Bell, J.F., Catalano, J.G., Clark, B.C., Crumpler, L.S., de Souza, P.A., Fairén, A.G., Farrand, W.H., Fox, V.K., Gellert, R., Ghosh, A., Golombek, M.P., Grotzinger, J.P., Guinness, E.A., Herkenhoff, K.E., Jolliff, B.L., Knoll, A.H., Li, R., McLennan, S.M., Ming, D.W., Mittlefehldt, D.W., Moore, J.M., Morris, R.V., Murchie, S.L., Parker, T.J., Paulsen, G., Rice, J.W., Ruff, S.W., Smith, M.D., Wolff, M.J., 2014. Ancient aqueous environments at Endeavour Crater. Mars. Sci. 343.

Baldock, T.E., Tomkins, M.R., Nielsen, P., Hughes, M.G., 2003. Settling velocity of sediments at high concentrations. Coastal Eng. 51, 10.

Balme, M.R., Bargery, A.S., Gallagher, C.J., Gupta, S. (Eds.), 2011. Martian Geomorphology. Geological Society, Special Publications, London, p. 356.

Barlow, N.G. (Ed.), 2008. Mars. An Introduction to its Interior, Surface and Atmosphere. Cambridge University Press, New York, p. 276.

Bell, J. (Ed.), 2008. The Martian Surface. Composition, Mineralogy, and Physical Properties. Cambridge University Press, New York, p. 652.

Cabrol, N.A., Grin, E. (Eds.), 2010. Lakes on Mars. Elsevier, Amsterdam, The Netherlands, p. 390.

Carr, M.H. (Ed.), 2006. The Surface of Mars. Cambridge University Press, New York, p. 322.

Julien, P.Y., 2010. Erosion and Sedimentation, Second Edition Cambridge University Press, Cambridge, p. 371.

Geiger, H., 2009. Astrobiologie. UTB, Stuttgart, p. 235.

Grotzinger, J.P., 2014. Habitability, Taphonomy, and the Search for Organic Carbon on Mars. Science 343, 4.

Grotzinger, J.P., Hayes, A.G., Lamb, M.P., McLennan, S.M., 2013. Sedimentary processes on Earth, Mars, Titan, and Venus. In: Mackwell, S.J. et al. (Ed.), Comparative Climatology of Terrestrial Planets. University of Arizona, Tucson, pp. 439–472.

Grotzinger, J.P., Milliken, R.E., 2012. The sedimentary rock record of Mars: distribution, origins, and global stratigraphy. In: Grotzinger, J.P., Milliken, R.E. (Eds.), Sedimentary Geology of Mars. Society of Sedimentary Geology Special Publication 102, Tulsa, Oklahoma, pp. 1–48.

Grotzinger, J.P., Sumner, D.Y., Kah, L.C., Stack, K., Gupta, S., Edgar, L., Rubin, D., Lewis, K., Schieber, J., Mangold, N., Milliken, R., Conrad, P.G., DesMarais, D., Farmer, J., Siebach, K., Calef, F., Hurowitz, J., McLennan, S.M., Ming, D., Vaniman, D., Crisp, J., Vasavada, A., Edgett, K.S., Malin, M., Blake, D., Gellert, R., Mahaffy, P., Wiens, R.C., Maurice, S., Grant, J.A., Wilson, S., Anderson, R.C., Beegle, L., Arvidson, R., Hallet, B., Sletten, R.S., Rice, M., Bell, J., Griffes, J., Ehlmann, B., Anderson, R.B., Bristow, T.F., Dietrich, W.E., Dromart, G., Eigenbrode, J., Fraeman, A., Hardgrove, C., Herkenhoff, K., Jandura, L., Kocurek, G., Lee, S., Leshin, L.A., Leveille, R., Limonadi, D., Maki, J., McCloskey, S., Meyer, M., Minitti, M., Newsom, H., Oehler, D., Okon, A., Palucis, M., Parker, T., Rowland, S., Schmidt, M., Squyres, S., Steele, A., Stolper, E., Summons, R., Treiman, A., Williams, R., Yingst, A., Team, M.S., 2014. A habitable fluvio-lacustrine environment at Yellowknife Bay, Gale Crater, Mars. Science 343, 14.

Knighton, D., 2014. Fluvial Forms and Processes: A New Perspective, Second Edition. Routledge, New York, pp. 383.

Komar, P.D., 1979. Modes of Sediment Transport in Channelized Water Flows with Ramifications to the Erosion of the Martian Outflow Channels. Icarus 42, 13.

Lamb, M.P., Grotzinger, J.P., Southard, J.B., Tosca, N.J., 2012. Were aqueous ripples on Mars formed by flowing brines? In: Grotzinger, J.P., Milliken, R.E. (Eds.), Sedimentary Geology of Mars. Society of Sedimentary Geology Special Publication 102, Tulsa, Oklahoma, pp. 139–150.

McEwen, A.S., Dundas, C.M., Mattson, S.S., Toigo, A.D., Ojha, L., Wray, J.J., Chojnacki, M., Byrne, S., Murchie, S.L., Thomas, N., 2014. Recurring slope lineae in equatorial regions of Mars. Nat. Geosci. 7, 53–58.

Melosh, H.J., 2011. Planetary Surface Processes. Cambridge University Press, Cambridge, UK, p. 500.

Vago, J., Gardini, B., Kminek, G., Baglioni, P., Gianfiglio, G., Santovincenzo, A., Bayón, S., van Winendael, M., 2006. Exo Mars. Searching for Life on the Red Planet. Report, 16–23.

INTERNET RESOURCES

http://exploration.esa.int/mars/

http://insight.jpl.nasa.gov/home.cfm

http://photojournal.jpl.nasa.gov

Mars Missions

Mission	Launch	Arrival at Mars	Termination	Elements	Outcome
USSR Mars 1M No. 1	October 10, 1960		October 10, 1960	Flyby	Launch failure
USSR Mars 1M No. 2	October 14, 1960		October 14, 1960	Flyby	Launch failure
USSR Mars 2MV-4 No. 1	October 24, 1962		October 24, 1962	Flyby	Broke up shortly after launch
National Aeronautics and Space Administration Mars 1	November 1, 1962		March 21, 1963	Flyby	Some data collected, but lost contact before reaching Mars, flyby at approx. 193,000 km
USSR Mars 2MV-3 No. 1	November 4, 1962		January 19, 1963	Lander	Failed to leave Earth's orbit
USA Mariner 3	November 5, 1964		November 5, 1964	Flyby	Failure during launch ruined trajectory
National Aeronautics and Space Administration Mariner 4	November 28, 1964	July 14, 1965	December 21, 1967	Flyby	Success (21 images returned)
Soviet Union Zond 2	November 30, 1964		May 1965	Flyby	Communication lost 3 months before reaching Mars
National Aeronautics and Space Administration Mariner 6	February 25, 1969	July 31, 1969	August 1969	Flyby	Success
National Aeronautics and Space Administration Mariner 7	March 27, 1969	August 5, 1969	August 1969	Flyby	Success
Soviet Union Mars 2M No. 521	March 27, 1969		March 27, 1969	Orbiter	Launch failure
Soviet Union Mars 2M No. 522	April 2, 1969		April 2, 1969	Orbiter	Launch failure
National Aeronautics and Space Administration Mariner 8	May 8, 1971		May 8, 1971	Orbiter	Launch failure
Soviet Union Kosmos 419	May 10, 1971		May 12, 1971	Orbiter	Launch failure

(Continued)

Experiments in Reduced Gravity: Sediment Settling on Mars. DOI: 10.1016/B978-0-12-799965-4.00013-3

Mission	Launch	Arrival at Mars	Termination	Elements	Outcome
National Aeronautics and Space Administration Mariner 9	May 30, 1971	November 13, 1971	October 27, 1972	Orbiter	Success (first successful orbit)
Soviet Union Mars 2	May 19, 1971	November 27, 1971	August 22, 1972	Orbiter	Success
			November 27, 1971	Lander, rover	Crashed on surface of Mars
Soviet Union Mars 3	May 28, 1971	December 2, 1971	August 22, 1972	Orbiter	Success
			December 2, 1971	Lander, rover	Partial success. First successful landing; landed softly but ceased transmission within 15 s
Soviet Union Mars 4	July 21, 1973	February 10, 1974	February 10, 1974	Orbiter	Could not enter orbit, made a close flyby
Soviet Union Mars 5	July 25, 1973	February 2, 1974	February 21, 1974	Orbiter	Partial success. Entered orbit and returned data, but failed within 9 days
Soviet Union Mars 6	August 5, 1973	March 12, 1974	March 12, 1974	Lander	Partial success. Data returned during descent but not after landing on Mars
Soviet Union Mars 7	August 9, 1973	March 9, 1974	March 9, 1974	Lander	Landing probe separated prematurely; entered heliocentric orbit
National Aeronautics and Space Administration Viking 1	August 20, 1975	July 20, 1976	August 17, 1980	Orbiter	Success
			November 13, 1982	Lander	Success
National Aeronautics and Space Administration Viking 2	September 9, 1975	September 3, 1976	July 25, 1978	Orbiter	Success
			April 11, 1980	Lander	Success
Soviet Union Phobos 1	July 7, 1988		September 2, 1988	Orbiter	Contact lost while en route to Mars
				Lander	Not deployed
Soviet Union Phobos 2	July 12, 1988	January 29, 1989	March 27, 1989	Orbiter	Partial success: entered orbit and returned some data. Contact lost just before deployment of landers
				Landers	Not deployed
National Aeronautics and Space Administration Mars Observer	September 25, 1992	August 24, 1993	August 21, 1993	Orbiter	Lost contact just before arrival

Mission	Launch	Arrival at Mars	Termination	Elements	Outcome
National Aeronautics and Space Administration Mars Global Surveyor	November 7, 1996	September 11, 1997	November 5, 2006	Orbiter	Success
Russia Mars 96	November 16, 1996		November 17, 1996	Orbiter, lander, penetrator	Launch failure
National Aeronautics and Space Administration Mars Pathfinder	December 4, 1996	July 4, 1997	September 27, 1997	Lander, rover	Success
Japan Nozomi (Planet-B)	July 4, 1998		December 9, 2003	Orbiter	Complications while en route; never entered orbit
National Aeronautics and Space Administration Mars Climate Orbiter	December 11, 1998	September 23, 1999	September 23, 1999	Orbiter	Crashed on surface due to metric-imperial mix-up
National Aeronautics and Space Administration Mars Polar Lander	January 3, 1999	December 3, 1999	December 3, 1999	Lander	Crash-landed on surface due to improper hardware testing
National Aeronautics and Space Administration Deep Space 2 (DS2)				Hard landers	
National Aeronautics and Space Administration 2001 Mars Odyssey	April 7, 2001	October 24, 2001	Currently operational	Orbiter	Success
European Space Agency Mars Express	June 2, 2003	December 25, 2003	Currently operational	Orbiter	Success
Great Britain Beagle 2			February 6, 2004	Lander	Landing failure; fate unknown
National Aeronautics and Space Administration MER-A Spirit	June 10, 2003	January 4, 2004	March 22, 2011	Rover	Success

(Continued)

Mission	Launch	Arrival at Mars	Termination	Elements	Outcome
National Aeronautics and Space Administration MER-B Opportunity	July 7, 2003	January 25, 2004	Currently operational	Rover	Success
European Space Agency Rosetta	March 2, 2004	February 25, 2007	Currently operational	Gravity assist en route to comet 67P/ Churyu- mov- Gera- simenko	Success
National Aeronautics and Space Administration Mars Reconnaissance Orbiter	August 12, 2005	March 10, 2006	Currently operational	Orbiter	Success
National Aeronautics and Space Administration Phoenix	August 4, 2007	May 25, 2008	November 10, 2008	Lander	Success
National Aeronautics and Space Administration Dawn	September 27, 2007	February 17, 2009	Currently operational	Gravity assist to Vesta	Success
Russian Federal Space Agency Fobos-Grunt	November 9, 2011		November 9, 2011	Phobos lander, sample return	Failed to leave Earth orbit. Fell back to Earth
China National Space Administration Yinghuo-1			November 9, 2011	Orbiter	
National Aeronautics and Space Administration MSL Curiosity	November 26, 2011	August 6, 2012	Currently operational	Rover	Success
Indian Space Research Organisation Mars Orbiter Mission	November 5, 2013	En route		Orbiter	Launched successfully
National Aeronautics and Space Administration MAVEN	November 18, 2013	En route		Orbiter	Launched successfully

Equations

The numbers correspond to the numbers in the text. When no unit is given, the factor has no unit.

$$q = w \times d \times v \qquad\qquad 1.1$$

q = runoff [m^3/s]
w = channel width [m]
d = depth [m]
v = velocity [m/s]

$$v = \frac{k_n}{nR^{2/3}S^{1/2}} \qquad\qquad 1.2$$

v = cross-sectional average velocity [m/s]
k_n = 1.0 for SI units
n = Manning coefficient of roughness
R = hydraulic radius [m]
S = slope of channel [m/m]

$$F_g = (\rho_p - \rho_f)g\frac{4}{3}\pi r^3 \qquad\qquad 4.1$$

F_g = gravitational force [N]
ρ_p = density of particle [g/cm^3]
ρ_f = density of fluid [g/cm^3]
r = radius of the particle [m]
g = gravity on earth or Mars [m/s^2]

$$F_d = 6\,\pi\mu r v_p \qquad\qquad 4.2$$

F_d = frictional force [N]
μ = kinematic viscosity [kg/ms]
r = radius of the particle in [m]
v_p = velocity of the particle [m/s]

Experiments in Reduced Gravity: Sediment Settling on Mars. DOI: 10.1016/B978-0-12-799965-4.00014-5

$$v_s = \frac{2(\rho_p - \rho_f)}{9\mu} gR^2$$ 4.3

v_s = terminal settling velocity [m/s]
ρ_p = density of particle [g/cm³]
ρ_f = density of fluid [g/cm³]
g = gravity on Earth or Mars [s⁻²]
R = differences between particle and fluid density [g/cm³]

$$w = \frac{RgD^2}{C_1 v_w}$$ 4.4

w = terminal velocity [m/s]
R = difference between particle and fluid density?
D = diameter [m]
C_1 = friction factor unit less
v_w = kinematic viscosity of water [kg/ms]

$$w = \sqrt{\frac{4RgD}{3C_2}}$$ 4.5

w = settling velocity [m/s]
C_2 = coefficient for turbulent drag, 0.4 for spherical particles, increasing with declining roundness
R = difference between particle and fluid density [g/cm³]

$$Re_p = \frac{U_p D}{v}$$ 4.6

Re_p = particle Reynolds number
U_p = velocity of the relative particle-fluid [m/s]
v = kinematic viscosity [kg/ms]
D = diameter [m]

$$C_d = \frac{F_d}{0.5\rho v^2 A}$$ 4.7

C_d = drag coefficient
F_d = frictional force [N]

A = reference area of the particle [m²]
ρ = density [g/cm³]
v = kinematic viscosity [kg/ms]

$$w = \frac{RgD^2}{C_1 v + \sqrt{0.75 C_2\, RgD^3}}$$

4.8

w = settling velocity [m/s]
R = difference between particle and fluid density?
D = diameter [m]
C_1 = friction factor
C_2 = coefficient for turbulent drag 0.4 for spherical particles, increasing with declining roundness
v = kinematic viscosity [kg/ms]

$$C_o = \frac{l_c}{(l_a \times l_b)^{0.5}}$$

4.9

C_o = Corey shape factor
$l_{a,b,c}$ = longest, intermediate, and shortest axis across a particle

$$C_d = \frac{24}{\mathrm{Re}} + \frac{0.5}{C_o^2}$$

4.10

C_d = drag coefficient
Re = Reynolds number

$$C_d = \frac{F_d}{0.5\rho v^2 A}$$

9.1

C_d = drag coefficient
F_d = frictional force [N]
A = reference area of the particle [m²]
ρ = density [g/cm³]
v = kinematic viscosity [kg/ms]

$$F_g = \sin(\alpha)mg \qquad\qquad 12.1$$

α = slope angle [°]
m = mass of water [kg]
g = gravity on Earth or Mars [s^{-2}]

$$v = \frac{k_n}{nR^{2/3}S^{1/2}} \qquad\qquad 12.2$$

v = cross-sectional average velocity [m/s]
k_n = 1.0 for SI units [m/s]
n = Manning coefficient of roughness unit less
R = hydraulic radius [m]
S = slope of channel [m]

$$ff = \frac{8gds}{v2} \qquad\qquad 12.3$$

ff = Darcy–Weisbach friction factor
g = gravity on earth or Mars [s^{-2}]
d = flow depth [m]
s = bed slope [m/m]
v = flow velocity [m/s]

$$\tau = \rho gdS \qquad\qquad 12.4$$

τ = shield parameter
ρ = density of the fluid [g/cm^3]
g = gravity on earth or Mars [s^{-2}]
d = flow depth [m]
S = slope of flow [m/m]

$$\tau^* = \Theta = \frac{\tau}{(\rho_s - \rho)gD} \qquad\qquad 12.5$$

τ^* = Θ = Shields parameter for a given sediment particle
τ = boundary shear stress [Pa]
ρ_s, ρ = density of the sediment and the fluid [g/cm^3]
g = gravity [m/s^2]
D = particle diameter [m]

Drag coefficients based on Reynolds numbers, $w = \log_{10} \mathrm{Re}$

For Re ≤ 0.01	$$C_{\mathrm{D}} = \frac{9}{2} + \frac{24}{\mathrm{Re}}$$
For 0.01 < Re ≤ 20	$$C_{\mathrm{D}} = \frac{24}{\mathrm{Re}}(1 + 0.1315\,\mathrm{Re}^{(0.82-0.05w)})$$
For 20 ≤ Re ≤ 260	$$C_{\mathrm{D}} = \frac{24}{\mathrm{Re}}(1 + 0.1935\,\mathrm{Re}^{(0.6305)})$$
For 260 ≤ Re ≤ 1.5 × 10³	$\log_{10}C_{\mathrm{D}} = 1.6435 - 1.1242w + 0.1558w^2$
For 1.5 × 10³ ≤ Re ≤ 1.2 × 10⁴	$\log_{10}C_{\mathrm{D}} = -2.4571 + 2.5558w - 0.9295w^2 + 0.1049w^3$
For 1.2 × 10⁴ ≤ Re ≤ 4.4 × 10⁴	$\log_{10}C_{\mathrm{D}} = -1.9181 + 0.6370w - 0.0636w^2$
For 4.4 × 10⁴ ≤ Re ≤ 3.38 × 10⁵	$\log_{10}C_{\mathrm{D}} = -4.3390 + 1.5809w - 0.1546w^2$
For 3.38 × 10⁵ ≤ Re ≤ 4 × 10⁵	$C_{\mathrm{D}} = 29.78 - 5.3w$
For 4 × 10⁵ ≤ Re ≤ 10⁶	$C_{\mathrm{D}} = 0.1w - 0.49$
For 10⁶ > Re	$$C_{\mathrm{D}} = 0.19 - \frac{8 \times 10^4}{\mathrm{Re}}$$

Research Proposal MarsSedEx II

1 RESEARCH PROPOSAL COVER PAGE

Principal Investigator:	Prof. Nikolaus J. Kuhn, PhD
Contact Name:	Prof. Nikolaus J. Kuhn, PhD
Contact Organization:	University of Basel
Mailing Address:	Physical Geography and Environmental Change, Department of Environmental Sciences, Klingelbergstrasse 27, 4056 Basel, Switzerland
Email:	nikolaus.kuhn@unibas.ch
Daytime Phone:	+41-61-267-0741
Alternate Phone:	+41-78-695-8403
Experiment Title:	Settling velocity of sand-sized quartz and basalt on Mars in water (MarsSedEx II)
Date Submitted:	September 20, 2013
Flight Date(s):	November 17, 2013
Flights Per Day:	One

Overall Assembly Weight (lbs.): approx. 25 pds
Stowed Dimensions (L × W × H): 15″ × 16″ × 21.5″
Deployed Dimensions (L × W × H): 15″ × 16″ × 21.5″
Equipment Orientation Requests: upright, i.e., the long side of the test apparatus upright
Proposed Floor Mounting Strategy (Bolts/Studs or Straps): University of Florida mounting plate with two belt strapping across the apparatus (see attached image of 2012 flight)
Power Requirement (Voltage and Current Required): None
Free Float Experiment: No
Number of Flyers Each Proposed Flight: One
Camera Pole and/or Video Support: Yes

2 TABLE OF CONTENTS

3 SAMPLE MANIFEST (ONE PER FLIGHT)

Flight Date: 11/18/2012

Flight Location: Titusville, FL

Name (as on government-issued, photo ID)	Birth date MM/DD/ YYYY	Male/ female	Prior parabolic flight exp (Y/N)	ZERO-G flight(s) – number or date	Identify experiment
Nikolaus Josef Kuhn	02/03/1970	Male	Yes	2, November 18 and 19, 2012	Sediment settling under Martian gravity (MarsSedEx II)

4 EXPERIMENT BACKGROUND

Aim of research

The experiment conducted for MarsSedEx II during the November 2013 flight aims at measuring the settling velocity of glass spheres, sand, and basalt particles of 0.5–1 mm diameter under Martian gravity in settling tubes filled with water resembling sedimentation in Martian runoff.

Particle settling velocity test in water

Particle settling velocity determines the time a particle requires to move through a liquid or gaseous medium from a given height to the base of the layer covered by the gas or liquid. If this liquid or gas is moving in a lateral direction, the settling velocity also determines the transport distance of the particle moving with the medium. On Mars, information on particle settling velocity is required to calculate the movement of loose particles in the atmosphere or running water. Such information can be used to assess environmental conditions, e.g., velocities of wind or rates of runoff required to form sedimentary rocks such as those currently analyzed by the Mars Rover Curiosity.

Settling velocities are determined by gravity, size, shape, and density of the particle, as well as density and dynamic viscosity of the gas or liquid the particle is moving in. When moving, the particle experiences friction and drag, which are both not independent of settling velocity. For simply shaped bodies, such as spheres, the settling velocities can be calculated using fluid dynamics or the empirical approach used in Stoke's law or its derivates, where the effect of friction and drag on settling velocity are represented by a reduction factor. For particles of a more complex, irregular shape, measurements are required which provide correction factors to account for the shape effects not covered by Stoke's or similar equations. Such simple correction works on Earth where gravity is constant. On Mars, simple corrections are unlikely to work because the aero- or aquadynamic environments at terminal velocity do not correspond to those on Earth. The results of the Mars-SedEx I flight in November 2012 confirmed this assumption because the estimated settling velocities were approximately 30% greater than those observed.

MarsSedEx II sediment settling velocity measurements

Particle settling velocity is commonly determined by measuring settling velocities in settling tubes. The MarsSedEx II experiment uses a set of four settling tubes developed from the MarsSedEx I design to measure settling velocities of reference glass spheres, quartz and basalt grains in water, simulating sedimentation on Mars. The results will help to assess flow hydraulic conditions under Martian gravity and the associated sediment settling. Overall, these tests will help to understand the geomorphic and climatic conditions required for erosion and deposition on Mars and thus contribute to our understanding of the environmental and ultimately potentially life-harboring scenarios on Mars.

5 EXPERIMENT DESCRIPTION

The MarsSedEx II particle settling experiment consists of four closed tubes (referred to as settling tubes) of 45-cm length and 5-cm diameter mounted in an upright position in a support structure sized 15″ L × 16″ W × 21.5″ H (Figure AIII.1). The tubes contain water and are closed to the outside environment all the time during the flight. The water has to be provided by Zero G for flight preparation on the day before the flight. At the top of each tube, two chambers separated by ball valves from the lower section of the tube contain either glass spheres, quartz, or basalt grains of 0.5–1 mm diameter. The volume of the water in each tube is 900 ml. At the beginning of the experiment, the top chambers of each tube are closed. Upon the onset of Martian gravity, the lower of the two chambers is opened so that the glass spheres, quartz, or basalt particles can settle through the water column toward the bottom of the tube. The second chamber is opened approximately 10 s later, enabling two

Fig. AIII.1. MarsSedEx II apparatus with four settling tubes with double ball valve chamber, cameras, and gravity logger.

measurements from each tube. The settling of the glass spheres, quartz, and basalt particles is recorded with video cameras mounted into the settling tube frame.

6 EXPERIMENT PROCEDURES DOCUMENTATION

All activities described below are conducted by the one researcher (Nikolaus J. Kuhn) taking part in the flight.

1. After take off, before Martian gravity parabolas: start cameras, clocks, video cameras, and gravity meter.
2. During first Martian gravity parabola: opening of two top chambers of two settling tubes containing glass spheres and quartz in water.
3. During second Martian gravity parabola: opening of top chambers of two settling remaining tubes containing quartz and basalt in water.
4. After parabolas: stop cameras, clocks, video cameras, and gravity meter.
5. After flight: remove apparatus from plane and dismantle.

7 EQUIPMENT DESCRIPTION

The MarsSedEx II settling tube apparatus is based on the design of the MarsSedEx I equipment flown onboard G-Force 1 in November 2012. The apparatus has the dimensions and weight prescribed by Zero G for our experiment (15″ L × 16″ W × 21.5″ H, 25 lbs). The support structure of the apparatus consists of a center square beam, screwed to a bottom and top plate. Four settling tubes are mounted each onto a side of the central square beam. Beam, plates, and clamps connecting tubes are manufactured from aluminum and designed to carry at least 10 times their weight in each direction. The same criteria were applied to the selection of the four bolts and threads that connect the plates to the central beam (Figure AIII.2). The tubes and ball valves can withstand/remain tight at an internal pressure of 16 bars, i.e., at least the eightfold pressure exerted by the liquid inside the tube at 2 g at the bottom of a parabola. The clamps holding the tubes and camera and their connections to support structure have been designed and tested to withstand a load of at least 10 times their own weight in each direction.

Fig. AIII.2. Top plate of the MarsSedEx II support structure with slots for strapping belts in each corner.

Further instruments used on the apparatus include:

1. an **MSR 145** gravity logger, attached with velcro to the frame
2. four GoPro and Hyundai action Camcorders, attached by clamps to the frame
3. an iphone or (ipad mini) used as g-meter, attached by velcro to the frame
4. a stopwatch to time the opening times of the sediment chambers, held by the **PI**
5. two digital watches, attached by velcro to the frame
6. two digital thermometers mounted in two settling tubes to record water temperature

The MarsSedEx II apparatus is strapped to the floor of G-Force 1 using the University of Florida mounting plate with two belt strapping across the apparatus similar to the 2012 MarsSedEx I flight (Figure AIII.2). The MarsSedEx II support structure top and base

Fig. AIII.3. Mounting of the MarsSedEx I apparatus during November 2012 flight.

plates have been designed accordingly and provide a diagonal slot (Figure AIII.3) in each corner for the strap belt.

The water in the settling tubes is regular tap water provided by Zero G. The glass spheres are produced by Micrometrics and were used already during the 2012 MarsSedEx I flight. The quartz and basalt grains have been sampled by N.J. Kuhn. All sediment particles are available for a safety inspection on the flight preparation day.

8 STRUCTURAL ANALYSIS

The MarsSedEx II support structures and settling tubes have been designed by a certified engineer (Dipl.-Ing. Wiedenhöft, Mageba Maschinenbau GmbH, Bernkastel-Kues, Germany) according to the standards prescribed by Zero G. The parts were built by Mageba Maschinenbau GmbH) and assembled by the Geowerk Basel (the University of Basel's workshop).

9 ELECTRICAL ANALYSIS

All electrical equipment (cameras, digital clock, g-meter, iphone) used for this experiment is commercially available and safe to carry and use on airplane. The equipment has not been changed from its original design.

10 HAZARD ANALYSIS REPORT GUIDELINES

None of the materials used for this experiment poses a safety or health hazard.

11 HAZARDOUS MATERIAL

No hazardous materials are used for this experiment.

12 MATERIAL SAFETY DATA SHEETS (MSDS) BIBLIOGRAPHY

N/A

MarsSedEx II Flight Plan

CRUISE AFTER TAKEOFF

1. Switch on and start Hyundai camera
2. Switch on MSR 145 by pressing start button and check for regular flashing of blue light
3. Switch on iphone (Code 6473) and open gmeter app
4. Check distance of camera 1 to tube on the RIGHT side, use short marked distance on the ruler

   Do not touch the button on the side of the camera.

5. Switch on camera 1: Press FRONT button and check for video camera symbol in front display using handheld mirror
6. Start recording by camera 1 by pressing TOP button
7. Repeat steps 4, 5, and 6 for cameras 2, 3, and 4
8. Start stopwatch and flash LEDs at the same time
9. Switch LEDs on
10. Note water temperature

PARABOLAS

1. Mars 1: Place left hand on the LOWER handle of opposite tube, wait for iphone gmeter to drop below 0.35, open valve by pressing on marked side; open top chamber 3 5 s later
2. Mars 2: Place hands on lower valve handles of tube 2 and 4, open when g is below 0.35, then open top chambers
3. Mars 3: Open bottom and top chamber of tube 1

   If running out of time, wait for Mars 4.

CRUISE AFTER PARABOLAS

1. Switch off cameras except for Hyundai, point toward yourself
2. Pack GoPro cameras, g-logger, and iphone, keep g-logger and gmeter app running
3. Record water temperature

Experiments in Reduced Gravity: Sediment Settling on Mars. DOI: 10.1016/B978-0-12-799965-4.00016-9